Where Vultures Feast

Where Vultures Feast

SHELL, HUMAN RIGHTS, AND OIL IN THE NIGER DELTA

IKE OKONTA
ORONTO DOUGLAS

VERSO
London • New York

Authors' note. In an effort to present as fair and balanced an account as possible, we wrote to officials of Shell Nigeria, requesting an interview to afford them an opportunity to tell us their own side of the story as they saw it, but our letter was unanswered. We, however, did not let this deter us. Royal/Dutch Shell and its Nigerian subsidiary, in the course of responding to charges and allegations that they had played a role in the exacerbation of the Ogoni crisis, have, since 1993, published briefing papers, memos, and official booklets explaining their position. We relied on these documents to offer the reader as balanced an account as the documents available to us would allow.

This edition first published by Verso 2003

First published by Sierra Club Books in conjunction with Crown Publishers, New York

1 3 5 7 9 10 8 6 4 2

Verso
UK: 6 Meard Street, London W1F 0EG
USA: 20 Jay Street, Suite 1010, Brooklyn, NY 11201

Verso is the imprint of New Left Books
www.versobooks.com

ISBN: 978-1-85984-473-1

British Library Cataloguing in Publication Data
A catalogue record for this book is available from the British Library

Library of Congress Cataloging-in-Publication Data
A catalog record for this book is available from the Library of Congress

Design by Leonard W. Henderson
Printed and bound in Great Britain by the Bath Press, Avon

To Nnah Uabari (aged nineteen),
who was murdered on October 25, 1993,
near Shell Flow Station No. 5, Korokoro, Ogoni, Nigeria

CONTENTS

ACKNOWLEDGMENTS

When we set out to tell the story of Shell in the Niger Delta in early 1996, we thought it would be a matter of gathering the evidence and putting it together in a book in a few months. We know now that telling the story of the biggest multinational oil company in the world is like traversing the world itself, an adventure that saw "a few months" stretching into three years of traveling, researching, "snooping," asking questions. It was a humbling experience.

We thank Nick Ashton-Jones and Nnimmo Bassey, whose deep knowledge of the human ecosystem of the Niger Delta and whose abiding love of the people proved an invaluable compass as we set out on our journey. We thank Ike Achebe, who would come down from his lodgings in Trinity College, Cambridge, read the manuscript in progress, and urge us on: "Facts, more facts!" We thank Andrew Rowell, Bronwen Mamby, Sheila Braithwaite, Chidi Anselm Odinkalu, Awa Dabo, and Jedrzeg George Frynas, who read drafts of the manuscript in 1997 and 1998 and offered very useful suggestions. We thank Owens Wiwa and Ken Wiwa, Jr. for putting aside their grief for a moment to offer us useful advice that made the telling of this story easier going.

Mary Isioma Arinze, Sarah Modebe, Chichi Iwedinwa, Ritje Grit, Urmi Shah, Nick Jukes, Mark Brown, Claudia Lehmkhul, Shlomi Segal, Nir Eyal, Glen Ellis, Kay Bishop, George Monbiot, Nana Yaa Mensah, Ebele Obumselu, Robert Beckford, Daphne Wysham, and Danny Moses, our very patient editor at Sierra Club Books, San Francisco, were there when it most mattered. Chima Ubani, Ogaga Ifowodo, Tajudeen Abdul-Raheem, Danbala Danju, Biyi Bandele, Makin Soyinka, Alaba Yusuf, Kayode Fayemi, and Andrew Chandler of the George Bell Institute took a personal interest in this project and encouraged us to see it through. ERA people, in Nigeria and all over the world, were, of course, the rock without whose solid support the project would have floundered. We thank Brixton, its people and warm ambience, for constantly reminding us, as we struggled to put this book together in a cold London bed-sit, that in the final analysis it is *people* that really matter.

We are grateful to Trócaire, the Catholic Agency for World Development, Dublin, for providing the financial assistance that made the researching and writing of this book possible.

Ike Okonta and Oronto Douglas
Niger Delta, Nigeria, July 1999

FOREWORD

Royal Dutch/Shell is more than a colonial force in Nigeria. A colonial power exhibits some measure of concern for the territory over which it lords. This is not the case with this mogul, which goes for crude oil in the most *crude* manner possible.

Four decades of oil production has led to major dislocations in the lives of the people of the oil-producing communities of Nigeria's Niger Delta. Violence done to their environment has translated into direct violence against the people. Ken Saro-Wiwa and the eight Ogoni patriots may well have been hanged on Shell's oil rig. Nothing is allowed to stand in Shell's way: not trees, not swamps, not beast, not man. The people of the Niger Delta have been forced to live with a highly polluted environment: the result of practices that would not be permitted in Europe or the United States.

Peace was banished the moment the first dynamite was exploded by Shell workers in Oloibiri village in search of oil. The situation became more precarious by the day, and the Niger Delta was the only part of Nigeria where a special military occupation force, set up by the federal government in 1994 and which had Shell's support at the time, took over the lives of the people, killing, maiming, and raping thousands.

We must pause and think again. If the experience of the people of the Niger Delta is anything to go by, the entire crude oil business is completely wrongheaded. There is not one stage of oil production that is sustainable or environmentally friendly. None. In Curacao, Shell, after operating a refinery for seventy years, packed its bags and left. Two asphalt lakes beside the refinery have turned pristine wetland in the area into wasteland. Clearly, there is worse to come in the Niger Delta, the world's most threatened human ecosystem.

This book unravels the true face of Shell and the hidden pains of a people mauled by Big Business and military dictatorship. Okonta and Douglas have given us an invaluable warning. Have you been warned?

Nnimmo Bassey
Director, Environmental Rights Action
Benin City, Nigeria

The wailing is for the fields of men:
For the barren wedded ones;
For perishing children . . .
The wailing is for the Great River:
Her pot-bellied watchers
Despoil her. . . .

Christopher Okigbo
"Lament of the Drums"

Where
Vultures
Feast

INTRODUCTION

On February 22, 1895, a British naval force under the command of Admiral Sir Frederick Bedford laid siege on Brass, the chief city of the Ijo people of Nembe in Nigeria's Niger Delta. After severe fighting, the city was razed to the ground. Over two thousand people, mostly women and children, perished in that attack launched in the name of Queen Victoria.

The 1895 massacre was at the behest of a British company, the Royal Niger Company, for which George Taubman Goldie had obtained a royal charter in 1886, giving it a monopoly of trade on the Niger River. Goldie, anxious to maximize his profits, moved right from the onset to displace merchants of Brass and the other surrounding communities who had acted as middlemen between the palm oil farmers in the hinterland and the European traders on the coast. He parceled off a vast area of land and imposed heavy duties on whoever might want to trade in his "territory"—thereby banning Brass traders from trading in their own land. After their several entreaties to the British Consul General were met with silence, the people of Brass took matters into their own hands and pulled down the Royal Niger Company trading post in Akassa. Taubman Goldie and the protectorate government's response to this attempt by the people of Nembe to safeguard their source of livelihood was the horrendous attack of February 1895—a military expedition in which the population of Brass and the nearby towns of Twon and Fishtown were almost wiped out.

One hundred years later, in February 1995, the people of Nembe were locked in a grim, death-and-life struggle with Royal Dutch Shell, another British firm, again to safeguard their source of livelihood: their environment, which the multinational oil company's exploration and production activities had despoiled. Shell, in collaboration with successive governments in Nigeria, has been extracting billions of dollars' worth of oil and gas from Nembe and other communities in the Niger Delta since 1956 without giving them much in return. The plunder of the Niger Delta has turned full cycle. Crude oil has taken the place of palm oil, but the dramatis personae are the same—a powerful European multinational company intent on extracting the last life juice out of the richly endowed Niger Delta, and a hapless people struggling valiantly against this juggernaut.

Shell, the operating company of the largest joint venture in Nigeria, accounts for some 50 percent of all oil production in the country, the bulk of it in the Niger Delta. The oil-producing communities therefore see Shell as the number-one culprit in the economic and ecological war currently being waged against them. Slowly but relentlessly, such oil production activities as gas flaring, oil spillage, indiscriminate construction of canals, and waste dumping have brought the human ecosystem of the Niger Delta to the point of near collapse. Shell acquires land and also degrades private property in the course of its oil production activities, and pays the affected communities little or no compensation. Nor do the communities receive a fair share of the oil royalties—the bulk of which is shared between the Nigerian government, Shell, and the other oil companies. Trapped between a vicious and morally bankrupt government and an unscrupulous multinational, these communities have now taken to the path of nonviolent protest in a bid to protect what little remains of their endangered environment and source of livelihood. *Where Vultures Feast* is the story of their encounter with one of the most powerful multinational companies in the world.

We begin our tale with a brief excursion to the past. Chapter One attempts to put the people of the Niger Delta and their environment in historical perspective and demonstrate that Shell is only the latest in the long list of robber barons that have plundered their land, beginning with the slave trade in the sixteenth century. This inhuman trade sucked the people of the Niger Delta into the orbit of international finance capital and, indeed, laid the basis for the exploitation of their resources by outsiders. Five hundred years later the pattern is unchanged—but Shell has added a frighteningly new dimension to this scenario: ecological warfare.

The gradual decay of Nigeria's political economy in the hands of an inept military and political elite—so preoccupied with plundering the oil wealth of the Niger Delta that they do not realize the country is dying—is the subject of Chapter Two. Royal Dutch Shell is one of the most profitable companies in the world.[1] A substantial part of this profit comes from the plum oil concessions it has garnered in the Niger Delta. The historical origins of the multinational and the manner by which it acquires and holds on to these lucrative oil fields in the Niger Delta and other countries is chronicled in Chapter Three. Profit is the engine that drives Royal Dutch Shell, and to maximize returns, rules have to be ignored and hapless "natives" deprived of what is theirs by right. Shell's ecological and economic war against the oil-producing communities of the Niger Delta, and its collaboration with successive Nigerian regimes to suppress them, is the focus of Chapters Four, Five, and Six.

The multinational also employs an army of public relations experts to maintain the image of a benevolent and environmentally friendly Big Brother in the Niger Delta, and this cynical game, played in the main by Shell spin doctors, is the subject of Chapter Seven. The alliance between the Nigerian military junta and the multinational to ignore and abuse laws and regulations guiding oil industry operations is examined in detail in the Appendix.

Ultimately, the question is: How do we heal the wound that Shell has inflicted, and continues to inflict, on the Niger Delta, one of the most endangered human ecosystems in the world? What must the international community, the people of the Niger Delta, and Nigerians at large do to stop this juggernaut from further damaging the area, threatening a people and their way of life?

Following the murder of Ken Saro-Wiwa—the author, environmentalist, and leader of the Movement for the Survival of the Ogoni People (MOSOP)—and eight of his compatriots by the military junta in November 1995, the political equation has changed. Nigeria's ethnic minorities are speaking out in a brave new voice and demanding that their wishes and aspirations be factored into the Nigerian project. The oil-producing communities are on the boil, and they have struck out for self-determination, insisting on a new Nigeria informed by true federalism, equity, justice, and negotiated cooperation. They are also insisting that Shell be called to account and compelled to pay reparations for despoiling their environment and taking away their mineral resources these past forty years, without paying them the royalties that are their just due. In the Niger Delta today, the struggle is for social and ecological justice.

This is a struggle that simply does not allow for "neutral" spectators. All must choose whose side they are on—Shell and the Nigerian government intent on holding the oil-producing communities of the Niger Delta down, or the victims who are struggling nonviolently to put an end to this tyranny. The struggle in the Niger Delta is also a struggle for the soul of Nigeria and its future. The late Nigerian scholar Professor Claude Ake made this clear in these words: "MOSOP and Ogoniland must survive and flourish for the sake of us all. For better or worse, MOSOP and Ogoniland are the conscience of this country. They have risen above our slave culture of silence. They have found courage to be free and they have evolved a political consciousness which denies power to rogues, hypocrites, fools, and bullies. For better for worse, Ogoniland carries our hopes. Battered and bleeding, it struggles on to realize our promise and to restore our dignity. If it falters, we die."[2]

ONE

A People and Their Environment

> *And finally, on the immense scale of humanity, there were racial hatreds, slavery, exploitation, and above all the bloodless genocide which consisted in the setting aside of fifteen thousand millions men.*
>
> Frantz Fanon
> *The Wretched of the Earth*

The Niger has the third-largest drainage area of Africa's rivers. The delta into which it drains is a huge floodplain in southeastern Nigeria consisting of sedimentary deposits flowing down from the Niger and the Benue rivers and covering 25,640 square kilometers of the country's total land area. This floodplain is home to some seven million people, grouped into several nations and ethnic groups: the Ijo, Urhobo, Itsekiri, Isoko, Efik, Etche, Ibibio, Igbo, Andoni, Ikwere, Ogoni, Isoko, Edo, and Kwale-Igbo. Some of the ethnic groups are further divided into clans with their own distinctive languages.[1]

Before the arrival of European traders in what is now modern Nigeria, the Niger Delta was inhabited mainly by the Ijo peoples, who lived in small creekside fishing villages ranging from two hundred to about a thousand inhabitants. The head of the village was the Amanyanabo (or Amakasowei), who in turn was elected by the heads of the various wards or patrilineages. With the advent of the slave trade, however, there was a rapid expansion of the population of the Delta. The hitherto small and idyllic Ijo fishing villages grew into powerful trading states like Bonny, Owome (New Calabar), Okrika, and Brass (Nembe), some of whose origins can be traced to the early sixteenth century. The Efik trading state of Old Calabar at the entrance of the Cross River, and the Itsekiri kingdom of Warri in the western Delta, also emerged at this time.[2]

The slave trade brought with it great social and economic upheavals in the Niger Delta.[3] Before the arrival of the European slave traders, the Ijo and the other peoples of the Delta traded with the peoples of the hinterland—mainly the Igbo and Ibibio. The former exported dried fish and salt to their neighbors in exchange for fruit and iron tools. The trade in slaves brought an abrupt stop to this flourishing commerce, however. The slave traders brought with them salt, dried fish, and new consumer goods such as cloth and metal utensils. The consumer goods were often cheap and not necessarily well made, but since the slave traders also brought salt along with them, the Ijo and the other inhabitants of the Delta gave up the trade in fish, salt, and iron tools with the Igbo and Ibibio altogether and concentrated on the lucrative slave trade.

It is generally assumed that the exploitation of the peoples of the Niger Delta and the devastation of their environment began when crude oil was discovered in the area by Royal Dutch Shell in 1956. The truth is that Europe's plunder of the Delta, and indeed the entire continent, dates much further back, to 1444, when the Portuguese adventurer and former tax collector, Lancarote de Freitas, sailed to the West African coast and stole 235 men and women whom he later sold as slaves.[4] De Freitas's trip was to trigger the Atlantic slave trade, which, before it was displaced by the trade in palm oil in the 1840s, saw several million able-bodied young men and women taken from the Delta and its hinterland and shipped to the plantations of North America, South America, and the West Indies.

The slave plantations of the West Indies were the basis of much British wealth. The Barclay brothers, David and Alexander, actively engaged in the slave trade in the 1750s and later used the proceeds to set up Barclays' Bank. William Gladstone's political career was funded by family wealth generated by his father's Liverpool trade and West Indies sugar plantations. In 1833, John Gladstone's assets included £296,000 (£15 million today) and £40,000 (£2 million)—or about $24 million and $3 million today—in Demerara and Jamaica respectively. William Gladstone's first speech in the House of Commons on June 3, 1834, was in opposition to the Slavery Abolition Bill, speaking as a West Indian representative. The staggering economic cost aside, slavery abruptly and catastrophically disrupted life in the Niger Delta and its hinterland, triggered interethnic wars, and led to the displacement of whole communities.

With the abolition of slavery in the first decades of the nineteenth century, there was a switch to the so-called "legitimate" trade in palm oil. But the pattern of trade remained unchanged—from the Niger Delta to Europe and back. Europe was at the height of its industrial revolution at this time, and the demand for palm oil, which was used to lubricate the machines of the factories and as raw material for soap and margarine, was high. The Delta traders played the role of middlemen between Liverpool merchants who anchored their ships on the coast and the cultivators of the palm oil in the hinterland.

At first this arrangement was satisfactory to all parties. Trade boomed. By 1850, British trading interests were concentrated mainly in Lagos, which provided access to the wealth of the forests of Yorubaland, farther west, and the Delta ports, which were the gateway to the interior of eastern Nigeria. Palm oil was now the chief export, as the European traders no longer found the trade in slaves profitable following the advent of the industrial revolution. Bonny, an Ijo town strategically located on the coast, gradually grew into the richest port in the Niger Delta, and by 1856 the port and its hinterland was exporting over 25,000 tons of palm oil a year, over half of the total quantity exported from Africa.[5]

Consuls and Gunboats

While the European slave merchants were content to ply their ignominious trade mainly from their ships using the kings and chiefs on the coast as go-betweens, the Liverpool palm oil barons began to actively interfere in the politics of the Niger Delta, beginning in 1850, with the sole purpose of displacing the local middlemen and appropriating the enormous profits for themselves. The argument of the Liverpool traders was that the middlemen of Bonny, New Calabar, Brass, and Old Calabar, the main Delta ports at the time, were not hardworking enough, and that the rich palm oil farms of the hinterland were not being exploited to the maximum as a result. They also complained about the high prices of the middlemen and increasingly began to urge the British government to intervene.[6]

Yet, as several chroniclers of trade and politics in the Niger Delta at the time have shown, it was actually the Liverpool merchants who were ripping off the coastal middlemen. They had a monopoly of the palm oil trade, and since there was not a standardized medium of exchange on the coast, they sold second-rate and sometimes worthless goods to the Africans. The historian K. O. Dike described trade practices at the time: "White supercargoes had managed to convince Africans that articles of clothing such as old soldiers' jackets and cocked hats bought at little cost at Monmouth Street, were a fair exchange for their raw materials."[7] Moreover, some of the British merchants were in reality ruffians and thieves and would actually seize barrels of palm oil from the Delta middlemen without payment.

It was to curb the activities of these rogue traders that the king of Bonny instituted a Court of Equity in 1854 run by a joint committee of the British traders and coastal middlemen under his supervision. Erring traders were fined, and those who refused to pay up were cut off from the palm oil trade by the Delta middlemen, who, obeying the king's directives, refused to sell palm oil to them. The Court of Equity brought order to the hitherto chaotic trade and was so successful that it was introduced in such other places as Akassa, Benin River (Itsekiri), Brass, and later Opobo.

The British merchants (or supercargoes) were, however, not satisfied. The palm oil trade was becoming even more lucrative as the pace of industrialization accelerated in Europe, requiring greater quantities of palm oil to lubricate machine parts and to manufacture soap and margarine for the millions of industrial workers who flocked to the cities. The supercargoes wanted a direct access to the hinterland so they could get the palm oil virtually for free. They began by using the British-appointed local consuls on the coast to force unfavorable terms of trade on the Delta middlemen. Indeed, the activities of the British consuls between 1850 and 1856 was to lead to the breakdown of the monopoly of coastal trade held by the supercargoes and the African middlemen between them. This, in turn, led to a crisis in Niger Delta politics.

John Beecroft, who was appointed Her Brittanic Majesty's Consul for the Bights of Benin and Biafra in 1849, laid the foundation of British power in Nigeria and initiated the politics that was to characterize the consular period in Nigerian history.[8] Beecroft saw himself not as an administrator but as a pathfinder of sorts, expanding British trade in the Niger Delta. It is instructive that the new consul's first intervention in the politics of the Delta was in palm-oil-rich Bonny, against King William Dappa Pepple, whom the supercargoes bitterly resented because they saw him as the main obstacle to their designs to get at the palm oil fields in the hinterland.

King Dappa Pepple had signed a treaty with Consul Beecroft in October 1850, regulating conditions of trade on the coast. In return, the British government had promised to pay the king an annual subsidy to enable him to develop the palm oil trade even further. But an increasingly powerful and ambitious Beecroft ignored the treaty to which he himself was a signatory, and even refused to pay the king the promised subsidy.[9] In 1851, Beecroft took the decisive first step—in what was to become his open intervention in Niger Delta politics—when he deposed Kosoko, the king of Lagos, and installed Akitoye in his place. Beecroft's excuse was that Kosoko was a slave trader, a practice the British government had decreed illegal. There was, however, ample evidence at that time to show that Akitoye was himself financed by a well-known slave trader, Domingo Jose, and certainly would have indulged in slave trading were it economically and politically expedient for him to do so. What Beecroft really wanted was a friendly king in Lagos who would help British merchants get a secure foothold in the area, and he conveniently used the "slave trader" tag to get rid of the independent-minded Kosoko.[10]

Beecroft employed similar tactics to do away with King Dappa Pepple of Bonny. He accused the king of sponsoring attacks on the ships of British traders on the New Calabar River, and, cleverly exploiting a trade dispute between Dappa Pepple and one of the royal lineages in Bonny, used the Court of Equity to deport him to Fernando Po in 1852.[11] After the removal of King Dappa Pepple, the British traders, in concert with the local consuls, accelerated the displacement of the Delta middlemen in the palm oil trade.

In 1855 some freed slaves from Sierra Leone who had converted to Christianity and settled in Calabar tried to help the local middlemen ship their palm oil directly to England, pointing out that the prices they got from the British supercargoes were ridiculous. The consul, Hutchinson, intervened, however, stopping the King, Eyo Honesty, when he tried to export a shipment directly to Liverpool. The consul claimed that the king owed £18,000 (about $30,000) to an English firm and so could not trade directly with the Liverpool commercial houses until he had paid it off.

A Commission of Inquiry later set up by the Foreign Office in London discovered that Hutchinson was corruptly enriching himself at the expense of the Niger Delta middlemen, and that he was in fact a commission agent in the employ of the English firm Hearn and Cuthbertson, to which he claimed the king of Calabar owed money. But this was after Hutchinson's predecessor, Consul Lynslager, had ransacked and destroyed the town of Old Calabar, claiming that he did so because the people practiced human sacrifice. Church of Scotland missionaries stationed in the town contradicted Lynslager, pointing out that the consul destroyed the town at the behest of British traders who wanted to teach the local middlemen a lesson for daring to trade directly with Liverpool.[12]

The enormous riches to be derived from the Niger Delta and the other coastal towns opened the eyes of the British traders and, subsequently, of the government itself, to the possibilities of taking over the area entirely, by force if necessary. Thus, in 1861, the Foreign Office instructed the consul to annex Lagos, "to protect and develop the important trade of which their town is the seat; and to exercise an influence on the surrounding tribes . . ."[13] Trade was growing by the day. The Niger provided an excellent highway for the British traders, who began to penetrate into the interior. They saw virgin forests brimming with agricultural produce. Fired by greed, they sent urgent dispatches to London. The Foreign Office, after ensuring that the area would not prove a financial liability to the government, but indeed the opposite, proclaimed the Niger Delta and its hinterland a British Protectorate in 1865, thus laying the foundations of what turned out to be modern Nigeria.

King Jaja and the Robber Barons

The story of King Jaja of Opobo and his epic struggle against the British merchants in the closing decades of the nineteenth century best illustrates the long-standing struggle of the peoples of the Niger Delta to protect their environment and its natural resources from the grasping hands of European mercantilists and their patrons in London, Paris, Hamburg, and Amsterdam.[14]

Jaja, who dominated the politics of the Niger Delta for twenty years, was an Igbo ex-slave in Bonny. Through hard work and a display of business acumen, he rapidly rose through the ranks and became head of the Anna Pepple royal house. Following a kingship tussle in the town, which escalated into civil war in 1869, Jaja and his followers retreated into Andoni country in the hinterland, named their new town Opobo, and declared it independent of the rulers of Bonny. Opobo, strategically located near the oil markets of the hinterland, quickly grew into the chief port in the Niger Delta, attracting European traders from all over the coast and even surpassing Bonny in wealth and political importance.

The British traders on the coast were, however, not happy with King Jaja. He had made it clear from the onset that he would not allow them direct access to the oil markets in the Opobo hinterland and that they could buy palm oil only from his agents. Jaja explained that since the British traders had a virtual monopoly over trade with the Liverpool commercial houses, he and his people should control the trade with the producers of the palm oil in the Delta hinterland.[15] New developments on the coast also favored King Jaja. In 1852 the British government had subsidized a fleet of steamers owned by Macgregor Laird, a merchant who began to operate a regular service between Liverpool and West Africa. This dealt a death blow to the great Liverpool houses and their monopoly of the Niger Delta oil trade. The Liverpool merchants began to face increasing competition, and by 1856 there were over two hundred European firms operating in the Niger Delta. Consul Lynslager's destruction of Old Calabar was a last-ditch attempt to prevent the local people from joining their European counterparts in turning this new development in shipping to commercial advantage. However, a few brave African middlemen began to export their oil directly to Europe, using Macgregor Laird's steamers. King Jaja, expectedly, was in the forefront.

This did not go down well with the British supercargoes, and they began to plot Jaja's downfall and to also devise means to evade his agents and buy palm oil directly from the hinterland. Jaja retaliated by increasing the volume of his shipments to England. Following a series of skirmishes with the supercargoes on the coast, King Jaja signed a treaty with the local consul in 1884 that effectively placed his town under British protection. But he made sure that a clause was inserted in the agreement that explicitly stated that his people would control the oil markets in the hinterland, his contention being that the supercargoes on the coast still controlled the bulk of shipments to England and took all the profits. Oil prices had, however, risen in Europe at this time, and the British supercargoes were getting increasingly impatient with Jaja. There were enormous profits to be made in the hinterland, and Jaja was in the way. They urged the local consul, H. H. Johnston, to intervene, and in 1887 King Jaja was deported to the West Indies. When he was eventually allowed to return to Opobo in 1891, he died on the way, a lonely, broken man.[16]

A similar fate was to befall Nana Olomu, a merchant prince and leader of the Itsekiri, who controlled the oil trade on the Benin River in the western Delta. Although Nana had signed a treaty with Consul Hewett in 1884, placing the Benin River, Warri, and some parts of western Ijo under British protection, he rebuffed attempts by the British to extend the powers of the new Oil Rivers Protectorate, which had been proclaimed in 1887, over his country. Nana correctly saw the new protectorate for what it really was: an attempt by the British traders on the coast to edge him out and take over the oil markets in his territory for themselves. But the supercargoes would brook no opposition. British gunboats were now in absolute control of the coast, the Niger River, and its tributaries. In September 1894, under the command of the acting consul, General Ralph Moor, they bombarded Nana's headquarters in Ebrohimie, ransacked the town, and carted away his goods.[17] Nana gave himself up a few months later. Thus was the last formidable obstacle to British imperialist designs in the Niger Delta removed.

Afterward, it was open season for the British merchants, most notably George Goldie Taubman, the "founder of modern Nigeria."[18] The scramble for Africa was going full steam when Goldie Taubman arrived in the Niger Delta. The French had set their eyes on the area, and Taubman decided that the only way to keep them out and secure the rich lands of the Niger basin for Britain was to wield the several British firms competing against one another in the Delta into a powerful trading bloc with total monopoly over the palm oil trade. The new firm that emerged from the merger was called the Niger Company. Taubman followed this up with a spate of "treaties" with the coastal kings, which he obtained literally at gunpoint. By 1884 he had obtained thirty-seven such "treaties." Towns that demurred, like Brass, Patani, and Asaba, farther inland, were bombarded into submission.[19]

When the conference of the European powers to divide Africa among themselves opened in Berlin in 1885, Taubman was the British government's official delegate. London was so pleased with his performance that his new company was granted a Royal Charter. In addition to the monopoly of the oil trade in the Niger districts that it already enjoyed, the company was given political authority over the area as well. Rechristened the Royal Niger Company, it set up its headquarters in Asaba.

The Birth of Nigeria

There is no doubt that George Goldie Taubman and the Royal Niger Company which he founded played a key role in bringing together the otherwise disparate nations and ethnic groups in the Niger Basin into what is now known as Nigeria. It must be pointed out, however, that Goldie Taubman's career as a monopoly trader in the Niger Delta and its hinterland was marked by looting, murder, and the mass sacking of whole towns and communities.[20] He was more a soldier than a trader—he came to the Niger Delta as a conqueror.

The palm oil wealth of the area provided Taubman the financial muscle with which the company now began to push forward into the hinterland, navigating the Niger up to Bussa in the north and opening new trading centers on its banks. Taubman, however, was not content with merely draining the Delta of its natural resources. He also embarked on trading practices that cut off the once flourishing Delta ports from the outside world, which plunged the populace into unprecedented penury from which it has never been able to recover. Indeed, it can be said that the basis of the underdevelopment of the Niger Delta, following the forcible integration of the area into the world of international finance capital in the nineteenth century, was laid by Goldie Taubman and the Royal Niger Company in the 1890s.[21]

Following the granting of a Royal Charter to the company in 1886, Taubman decreed that such towns as Brass, the chief port of the Nembe Ijo, which were outside the Royal Niger Company's territory, were "foreigners." The people of Brass were therefore forced to pay fifty pounds (about eighty dollars) a year for a trading license and another ten pounds for each company station they traded in. Trade in alcoholic spirits had also grown in importance, and Taubman imposed an extra hundred pounds (equivalent to about $8,000 today) tax on any Brass merchant who desired to trade in the commodity.[22] In the 1880s this was a lot of money. Perhaps the people of Brass might have endured this hardship in silence if the Royal Niger Company had not taken its monopoly practices further by preventing them from shipping their palm oil directly to England, insisting that all such exports be routed through Akassa, the company's port. The company also undercut the Brass traders by journeying into the hinterland to buy palm oil directly from the producers.

Faced with increasing poverty and hardship, the Brassmen revolted. In 1895 they attacked the company's port at Akassa and took sixty-seven men hostage, insisting that they would not be released until the company gave them access to their old markets in the hinterland. The Consul General of the newly established Niger Coast Protectorate sent a naval force to Nembe Creek, attacked the town, and razed it to the ground.[23] Two thousand unarmed people, mostly women and children, were murdered. The ancient kingdom of Benin farther west was to be subdued and brought under British rule two years later.

In 1894, Captain Frederick Lugard, a veteran of the East African campaign, arrived to help extend the British empire on the Niger coast, accelerating the pace of the "pacification" of the peoples of the Niger basin. The French were pushing aggressively from the Borgu area north of the Benue River, grabbing whatever territory they could lay hands on. In April 1898, London was sufficiently worried to ask Lugard to set up an armed unit to protect all the territory then under the control of the Royal Niger Company. Thus was born the West African Frontier Force, a battalion of soldiers that Frederick Lugard used to bring the vast Sokoto Caliphate under British control a few years later.[24]

In 1898, London withdrew the charter from the Royal Niger Company and set up structures to administer direct imperial control on her new domain. The palm oil trade alone was worth almost £3.4 million a year to Britain (£175 million today, or $280 million), and it was felt that the Niger basin was too valuable to be left to a commercial firm to manage. West of the Niger, Sir Gilbert Carter, the governor of Lagos, had by now brought the Yoruba states and kingdoms, which had been at war with one another for a hundred years, under British suzerainty. During his trek of 1893, Carter had used a mix of diplomatic cunning and force of arms to subdue the states and kingdoms.

On January 1, 1900, Britain's new domain was restructured under three administrative zones: the Protectorate of Southern Nigeria; the Lagos Colony; and the Protectorate of Northern Nigeria. The name "Nigeria" was chosen for it. It was left to Lugard to bring the Sokoto Caliphate to heel, and this he achieved between 1900 and 1906. The Aro, the only remaining obstacle to British colonialism in the Igbo heartland in the east, was conquered in 1901, but it took a little longer—well into the 1920s—before the entire area was "pacified" and brought under British rule. In 1914 the southern and northern protectorates were amalgamated under a single administrative unit and a new country was born.[25]

One Country, Many Nations

Nigeria, it must be remembered, began life as a loose collection of nations, ethnic groups, clans, and villages brought together under one roof by British force of arms. As the late politician Obafemi Awolowo put it, "It was [the British] who created Nigeria out of a welter of independent and warring villages, towns, and communities, and imbued the various Nigerian national groups with an overriding desire for the unity of the entire Federation."[26]

Before the 1914 amalgamation, Nigeria consisted of two distinct colonial territories, separately ruled and administered. The Sokoto Caliphate, founded by the Islamic warrior and scholar Uthman Dan Fodio in 1817, was a theocratic state—at least in theory—and was administered along lines outlined by its founder in his book *Kitab al-Farq*. The Caliphate was more closely linked to North Africa and Saudi Arabia, culturally and commercially, than with its neighbors west and east of the Niger, who were mainly Christians and traditional religion practitioners and had had contact with Europe dating back to the fifteenth century.

The interests of British trade were paramount, however, and the dictates of commerce, coupled with the financial difficulties of administering the various nations and ethnic groups as separate entities, compelled the colonial administrators, from Frederick Lugard onward, to treat the country as a single unit, using a system of "indirect rule" in the North and "direct rule" in the South. While the northern emirs who held unchallenged sway over their subjects were allowed to administer their territories with minimal interference from the colonial residents, Lugard discovered that this system of indirect administration could not apply in the more egalitarian south, where the ruler's authority was circumscribed by a large number of checks and balances. The South was therefore ruled directly through courts and a "warrant" system whereby certain individuals were raised to positions of authority specifically to dispense justice and collect taxes as the emirs did in the North.[27] The British were, however, determined to rule the country as two separate political units, employing the infamous tactics of divide-and-rule that they had perfected in India to keep the various indigenous groups constantly at each other's throats.

The 1922 Constitution, introduced by Lugard's successor as Governor General, Sir Hugh Clifford, provided for the first time for elected African members in a legislative council. The 1930s and early 1940s witnessed rapid social and political changes in colonial Nigeria. The Eastern and Western regions were created out of the old Southern Nigeria by administrative fiat in 1939, while the Northern Region was left intact. A small but educated and articulate indigenous elite had emerged. The Second World War also saw Nigerian soldiers serving alongside their European counterparts on an equal footing, and this further accelerated political consciousness among the population. Led by Western-educated journalists and politicians, they began to agitate for greater participation in the administration of the country. The Richards Constitution, which became effective in January 1947, was the colonial government's attempt to accommodate the demands of the nationalists by attempting "to secure greater participation by Africans in the discussion of their own affairs."[28] Governor Arthur Richards's constitution united the northern and southern parts of the country in one central legislature for the first time. Richards, though, made provisions for regional councils, thus ensuring that the North enjoyed a degree of autonomy and was not "contaminated" by the southern politicians, whom the colonialists generally looked down upon as upstarts and political agitators. The Richards Constitution thus helped lay the foundation of tribalism in Nigerian politics and proved a most effective counterfoil to the nationalistic, pan-Nigerian outlook of the National Council of Nigeria and Cameroons, which Dr. Nnamdi Azikiwe founded in August 1944, with the aim of driving the colonialists from the country.[29]

Arthur Richards also established the basis for an unequal and unwieldy federation, with the northern region twice the size of the East and West. Like his predecessors, Richards refused to listen to wise counsel from C. L. Temple, Lugard's lieutenant governor of the North, and restructure the country into seven or eight provinces, generally corresponding with the geographical space occupied by the various ethnic nationalities.[30]

Richards had hoped that his constitution would last for nine years. But this was not to be. As soon as a new governor, Sir John Macpherson, was appointed in April 1948, he announced that he would give the country a new constitution that would further widen the participation of the people in the political process. The Macpherson Constitution, which replaced Richards's in January 1952, put in place a federation with a central legislature and executive, but at the same time the regional assemblies were enlarged and given legislative and financial powers. Perhaps this was Macpherson's attempt to widen the democratic space at the regional level. The North, however, was given half of the seats in the central legislature of 148 members, ensuring its near total dominance of the nascent country's politics. It is instructive that such regional and ethnic-inspired political parties as the Action Group and Northern Peoples Congress (NPC) emerged in the West and North respectively, at this time.

The series of constitutional conferences that were later to culminate in independence for the country on October 1, 1960, under an NPC-led government were attempts by Nigeria's political leaders to fashion a federal constitution that would work smoothly "to promote efficiency in, and harmonious relations and unity among, the constituent parts of the Federation."[31] This laudable goal proved difficult to achieve, however, partly due to the lopsided nature of the federation the British left behind, and partly due to corruption, intolerance, and abuse of office on the part of the politicians. The breakdown of law and order in the Western Region in late 1965, orchestrated by the NPC—which wanted to crush the Action Group, the party of the opposition—triggered a chain of events culminating in a military coup led by Chukwuma Kaduna Nzeogwu, a young major, in January 1966. Several politicians and military officers were killed, among them the Prime Minister, Alhaji Abubakar Tafawa Balewa.

Attempts by the new military Head of State, General Johnson Aguiyi-Ironsi, to introduce a new unitary constitution in May 1966 sparked a mutiny by young Hausa-Fulani military officers, who claimed that the January coup was an attempt by the Igbo to take over the political leadership of the country. Ironsi was killed in a countercoup in July, and the killing of Igbos and other easterners began in northern towns and cities. There was a massive exodus of the latter to the Eastern Region, and when Ironsi's successor, Colonel Yakubu Gowon, proved incapable of stopping the genocide in the North, the military governor of the East, Colonel Emeka Odumegwu-Ojukwu, called on his fellow easterners in other parts of the country to return home. A new constitutional arrangement making for a loose confederation of the three regions was worked out for the country by the two sides in Aburi, Ghana, as a last-ditch effort to stave off civil war and the subsequent disintegration of the country. But Gowon reneged on this agreement after realizing that the Aburi Accord effectively gave the Eastern Region political autonomy. On May 27, 1967, he announced that the country would henceforth be divided into twelve states. Ojukwu saw this as an attempt to bury the Aburi Accord, and he responded three days later by proclaiming the former Eastern Region as the sovereign Republic of Biafra. The federal government declared war on Biafra on July 6, a bloody carnage that did not stop until Nigerian troops forcibly brought the East back into the federation in January 1970. An estimated two million people, the bulk of them Biafran children, lost their lives in this conflagration.

People of the Niger Delta Today

The bulk of the inhabitants of the Niger Delta live in three states in present-day Nigeria—Rivers, Delta, and the newly created Bayelsa. These states take up about 80 percent of the area. The rest are scattered in such other states as Cross Rivers, Akwa Ibom, Imo, and Ondo.

The years of slavery, palm oil trade, and subsequent colonial conquest brought with them massive migrations and intermingling of ethnic groups in the Delta. The rapid growth of Port Harcourt, the area's biggest city, in the decades leading to independence also encouraged intermarriage and resettlement of whole communities. As a result, today the Niger Delta is a fascinating collage of ethnic nationalities, clans, and language groups that, while still relatively distinct, nevertheless have many cultural similarities.

The Niger Delta has substantial oil and gas reserves. Oil mined in the area accounts for 95 percent of the country's foreign exchange earnings and about one-fourth of Gross Domestic Product. The bulk of Nigeria's proven oil reserves, currently estimated at twenty billion barrels, is located in the area, although exploration is also going on in the state of Bauchi and the Lake Chad Basin. Rivers, Bayelsa, and Delta states alone currently produce three-fourths of the country's crude oil.[32] Besides its great mineral wealth, the Niger Delta also has fertile agricultural land, forests, rivers, creeks, and coastal waters teeming with fish and sundry water creatures. Clearly, the Niger Delta is, at least for the moment, the goose that lays Nigeria's golden egg.

Yet, in spite of its considerable natural resources, the area is one of the poorest and most underdeveloped parts of the country. Seventy percent of the inhabitants still live a rural, subsistent existence characterized by a total absence of such basic facilities as electricity, pipe-borne water, hospitals, proper housing, and motorable roads. They are weighed down by debilitating poverty, malnutrition, and disease. While decades of corruption and mismanagement in the echelons of power have plunged the country's GNP per capita to an all-time low of $280, annual incomes in the Niger Delta are still far below the national average. The area also has one of the highest population densities in the world, and annual population growth is currently estimated at 3 percent. Rapid population growth is increasingly exerting pressure on cultivable land, a good part of which is in any case prone to flooding almost all year. The population of Port Harcourt and the other major towns is literally exploding. The ensuing scenario—urbanization without the economic growth that would ordinarily generate more jobs—has resulted in the human ecologist's ultimate nightmare: a growing population that, in a bid to survive, is destroying the very ecosystem that should guarantee its survival.

Historically, the people of the Niger Delta have always been at the mercy of greedy outsiders who plunder their natural resources without giving them anything in return, from the days of slavery to the present day. The civil war, however, was a watershed in the political and economic development of the peoples of the Niger Delta. It created the conditions for the accelerated exploitation of their resources and the devastation of their environment. Following the takeover of the Shell oil terminal in Bonny from Biafran troops, the Gowon regime enacted the Petroleum Act in 1969, which transferred all oil revenue to the Federal Military Government, which in turn was expected to disburse the money to the various states, partly on the basis of need. This decree and the subsequent legislation enacted by the military government in Lagos was to transform Nigeria from a genuine federation to a de facto unitary state.

But the unitary state that General Gowon and his military-dominated cabinet imposed on Nigerians did not have room for the peculiar needs of the minority people of the Niger Delta, even though they produced all the oil. The revenue went straight to the coffers of the Federal Military Government, the bulk of which was spent to finance the expensive lifestyle of the indolent and unproductive elite. Oil revenue and the corruption it engendered brought great wealth to this parasitic economic class. By 1976, Nigeria had become the seventh-largest producer of oil in the world, exporting two million barrels of crude a day. Federal revenue had risen to a staggering $5 billion per annum, compared to $590 million in 1965. The oil boom was in full swing, and members of the political and economic elite began to live it up, acquiring an unrivaled taste for imported Western luxuries.

TWO

Soldiers, Gangsters, and Oil

It can't be the massive corruption, though its scale and pervasiveness are truly intolerable; it isn't the subservience to foreign manipulation, degrading as it is . . . It is the failure of our rulers to reestablish vital links with the poor and dispossessed of this country.

Chinua Achebe
Anthills of the Savannah[1]

"To Keep Nigeria One Is a Task That Must Be Done"

Oil is the stuff of contemporary Nigerian politics, and the Niger Delta is the field on which the vicious battle to control this money spinner is waged. The civil war that raged between the breakaway Eastern Region and the rest of the country from July 1967 to January 1970 was not so much a war to maintain the unity and integrity of the country (fought with such catchy slogans as "To Keep Nigeria One Is a Task That Must Be Done") as a desperate gambit by the federal government to win back the oil fields of the Niger Delta from Biafra. The July 29, 1966, counter-coup led by young northern officers in the army was driven, first, by a desire to avenge the northern political leaders who died in the Major Kaduna Nzeogwu-led coup of January 15, 1966, but more important, was meant to lay the groundwork for the secession of the then Northern region from the rest of the country. These officers claimed that the Nzeogwu coup was ethnic-motivated.[2] In order to facilitate the secession of the North from the federation, they hunted down and killed the Head of State, General Johnson Aguiyi-Ironsi, an Igbo who had taken over the reins of government after the First Republic collapsed.

The coupists, however, changed their tune when the matter of oil came up. Lieutenant Colonel Yakubu Gowon, a northerner who effectively became Head of State after General Ironsi's death, was alerted to the danger of letting the Eastern Region go, with its rich oil reserves; suddenly he was no longer prepared to listen to the suggestion of eastern leaders, after thousands of their people had been slaughtered in northern towns in the wake of the July countercoup, that a confederal arrangement be worked out for the country to allow tempers to cool on both sides. As Sarah Khan, the petroleum economist, has written, "The renunciation of the [Aburi] agreement was related mainly to the issue of oil revenue distribution. The fear that the Eastern Region would, in fact, benefit greatly from partial autonomy and therefore greater control over its substantial oil wealth, gave momentum to the degeneration of affairs into civil war."[3]

It is instructive that the first legislation the new Head of State passed as soon as the Bonny terminal and the other oil fields of the Niger Delta were "liberated" from Biafra was the Petroleum Decree No. 51 of 1969, transferring all oil mineral rights and revenue accruing therefrom to the Federal Military Government. The eastern part of the country, and particularly the Niger Delta and its inhabitants, have been treated as conquered territory ever since. The poet and playwright Wole Soyinka made reference to this in his play *Opera Wonyosi*:

> *"Secession" cried one; the other "One Nation"*
> *For oil is sweet, awoof no get bone*
> *The task was done, the nation is one*
> *We know who won and who got undone*
> *No thought of keeping his body one*
> *It's scattered from Bendel to Bonny Town.*[4]

Shell had teamed up with British Petroleum, partly owned by the British government, to open up the Nigerian oil fields. Following the discovery of oil in Oloibiri village in the eastern Delta in 1956, the joint venture had begun to produce some 367,000 barrels of oil per day by the time the civil war broke out in July 1967.[5] The problem for senior Shell officials was not the rights and wrongs of the Biafran cause, but how to ensure that the federal government, which enjoyed the support of the British government, now part owner of the company, won the war. The way Shell saw it, only a federal victory could guarantee its continued position in the Nigerian oil industry, which it completely dominated at the time. On its own part, the Harold Wilson government chose to ignore the moral issues raised by Biafra's quest for self-determination following the slaughter of easterners in the former Northern Region, and supplied the Gowon-led federal government the arms and logistics with which Biafra's secession bid was defeated. As one Commonwealth Office briefing document in some of the Cabinet papers released in London in December 1997 made clear, "the sole immediate British interest is to bring the [Nigerian] economy back to a condition in which our substantial trade and investment can be further developed."[6]

Britain's interest in Nigeria's oil dates back to the foundation of the country itself. The Colonial Mineral Ordinance, enacted by Frederick Lugard shortly after he amalgamated Northern and Southern Nigeria in 1914, was the first oil-related legislation in the country.[7] The 1914 ordinance made oil prospecting in the new country a British monopoly with ownership rights vested in the crown. The 1937 Colonial Mineral Ordinance gave Shell D'Arcy (Shell's operating name in Nigeria at the time) exclusive exploration and prospecting rights in the country, and the Colonial Office followed this up a year later with a grant of an Oil Exploration License to the company covering the entire country. After Shell began oil production from its Oloibiri well in 1958, the colonial government enacted the 1959 Petroleum Profits Tax Ordinance, putting in place a fifty-fifty profit-sharing arrangement between the Nigerian government and foreign oil companies. Instructively, this was shortly before Nigeria gained independence in October 1960. Gowon's Petroleum Act of 1969, however, not only annulled the 1937 ordinance by transferring ownership of oil mineral rights to the federal government, it also overturned the country's revenue allocation formula, wherein the three regions had a fifty-fifty revenue-sharing agreement, in favor of the central government in Lagos, following the dissolution of the three regions by decree in 1967.[8]

The civil war was a watershed in the political and economic development of the peoples of the Niger Delta, creating the conditions for accelerated exploitation of their resources and the devastation of their environment. Although Shell had begun to produce a substantial amount of oil in the Niger Delta by the early 1960s, its contribution to national revenue was still negligible. At independence in 1960, Nigeria was virtually self-sufficient in food, and agricultural products accounted for 97 percent of export revenue. Things were to take a dramatic turn in the 1970s, however, as international oil prices began to rise. From a modest $295 million in 1965, federal revenue surged to $2.5 billion a mere ten years later. Oil accounted for 82 percent of this new wealth.[9] Nigeria had joined OPEC and was now in the front rank of oil-producing nations.

Very little of this wealth found its way to the communities of the Niger Delta from whose land the oil was extracted, however. Prior to the civil war, revenue allocation was based on the principle of derivation and largely devolved on the various regions, which were required by Section 134(1) of the 1960 constitution to give 50 percent of royalties and mining rents to the federal government. The Republican Constitution of 1963, under Section 140(1), retained these provisions.[10] The civil war and its aftermath changed all this. The leaders of the new military government were soldiers who had been trained and had pursued their careers in a unified and centrally controlled administration. They therefore found the federal system with its several autonomous units, each with its respective powers, irksome. There were also more compelling reasons why Gowon and his team jettisoned the old revenue allocation formula.

Although two new states, Rivers and South Eastern, had been carved out of the now defunct Eastern Region a few days before the outbreak of the civil war in July 1967, ostensibly for the benefit of the people of the Niger Delta, the inhabitants saw little of the country's newfound oil wealth. It needs to be stressed, however, that long before oil was discovered in the area in 1956, the ethnic groups that inhabit the Delta and other parts of southeastern Nigeria had regularly expressed fears of ethnic domination and discrimination in jobs and development assistance. Indeed, in an attempt to remedy the situation, political leaders of the area began to agitate for a new state to be carved out for them distinct from the core Igbo areas in the then Eastern Region in the fifties. The Willink Commission, set up by the 1957 constitutional conference, toured the minority areas of the country, and while it accepted that the grievances of the Niger Delta inhabitants were genuine, declined to recommend the creation of a new state for them, insisting that their interests would be protected in a new federal constitution in which fundamental human rights for minority peoples would be entrenched.[11]

With the new 1969 Petroleum Decree enacted by the victorious federal government firmly in place and oil prices increasingly on the rise, the new military leaders found it expedient—for obvious reasons—to discard the revenue allocation formula agreed to by the three regions in 1954, dividing mining revenue equally between the regions and the federal government. The government began to disburse money to the states partly based on need and partly on derivation. In theory, Nigeria was still a federal republic, but in day-to-day administration it was a de facto unitary state, run by a Supreme Military Council, with military governors in the twelve "provinces" accountable to it in every material particular. As agricultural exports in the various states declined and oil production soared, putting billions of dollars into federal coffers, the Federal Military Government became even more powerful, spending the oil money as it wished. The Niger Delta, still regarded as part of "conquered Biafra," was ignored. The 1996 Greenpeace report on social and environmental conditions in the Niger Delta described the situation thus: "For years Nigeria's oil-rich southeast has been considered a colony by the Nigerian political and military elite. Those oil revenues that the elite did not use to acquire luxury goods and top up foreign bank accounts, were spent exclusively in the north or southwest."[12]

It took Decree 6 of 1975, which increased the federal government's share of the oil proceeds from 50 to 80 percent, leaving the states with only 20 percent, to formalize this blatant injustice that had been going on since Gowon passed the Petroleum Decree in 1969. Even then, the communities of the Niger Delta that were dispersed in three states at the time—Rivers, Midwest, and South Eastern—saw very little of this "20 percent." While the elite in the cities lived it up during the heady oil boom years, peasants in the Niger Delta were facing starvation as a result of the drought of 1972-74, which drastically reduced agricultural production. The oil boom and the salary increases in the Federal Public Service in April 1974 had triggered an inflationary spiral in the country, but there was not a corresponding increase in the price of such export crops as palm oil and rubber, on which the inhabitants depended for extra income. There was no one to whom they could turn to for help. In any case, one of General Gowon's senior permanent secretaries had cynically remarked in a public lecture that the people of the Niger Delta were most unlikely to pose any real threat to the regime's continued exploitation of their oil wealth, as they were relatively few in population and thus could easily be subdued.[13] And this is exactly what the Nigerian military junta has done, beginning from the mid-eighties when the people of the Niger Delta began to raise their voices in protest.

Securing the Booty

Despite these assurances, there was still a certain amount of unease among the military and civilian elite. Their worst nightmare was another Biafra springing up in the Niger Delta, whose oil fields were at this time accounting for 82 percent of national revenue. In 1978, one year before the military government handed over power to the civilian regime of Alhaji Shehu Shagari, the soldiers passed the Land Use Act (originally Land Use Decree), vesting ownership and control of all land in the military governors of the various states as representatives of the Federal Military Government.[14] Prior to this decree, land was communally owned and the various traditional rulers, clan heads, and community leaders had the power to determine customary law insofar as this affected land tenure and land use. The new Second Republic constitution, whose drafting was supervised by the outgoing military government, also ensured that the people of the Niger Delta could not challenge the expropriation of their natural resources by the central government in Lagos in a court of law. Section 40(3) of the 1979 constitution vested control of minerals, mineral oil, and natural gas in, under, and upon any land in Nigeria in the federal government, and also conferred upon the National Assembly the power to make laws regarding revenue allocation.[15]

Thus, by the time the soldiers handed power back to the politicians in October 1979, twelve years after the first bullets were fired in the battle for the oil fields of the Niger Delta, the area had been secured and made safe for the ruling elite, who had come to regard the billions of dollars' worth of oil extracted every year as theirs by right. The only notable law the politicians of the short-lived Second Republic, under the leadership of President Shehu Shagari, passed with regard to sharing the oil revenue was the Allocation of Revenue Act No.1 of 1982, amended by the military regime of General Muhammadu Buhari two years later through Decree 36 of 1984. This act granted 55 percent of the oil revenue to the federal government (down from 80 percent in 1975), 32.5 percent to the states, and 10 percent to the local governments. One percent of the oil proceeds was set aside for disaster relief, and another 1.5 percent was established as a special fund for the oil-producing areas.[16]

It must be pointed out, however, that all through the Second Republic—whose checkered, corrupt-ridden life was abruptly terminated in December 1983 by General Buhari and other senior military officers—administrative incompetence, corruption, and interparty rivalry ensured that the special fund set aside for the communities of the Niger Delta was not translated into concrete social and economic benefits for the inhabitants. With the return of the military to power in 1983, the country reverted to a de facto unitary system of government once again, and whatever hopes the people of the Niger Delta harbored of pressing for an increase in the special fund allocated to them, and for a new law putting these monies under their direct control, evaporated. General Ibrahim Babangida, Buhari's Chief of Army Staff, ousted his boss in a palace coup on August 27, 1985, and beginning in 1986 unleashed a vicious and debilitating economic war on the populace with the collaboration of the International Monetary Fund (IMF).[17]

"This Second-Class, Hand-Me-Down Capitalism"[18]

As we argued in Chapter One, Nigeria was created by British merchants and soldiers of fortune primarily to serve the mother country's interests as nineteenth-century capitalism entered the stage of imperialism, and desired even more sources of cheap raw material and also new markets for its products. Thus, Lugard and his successors were not interested in putting in place an integrated and balanced economy that would benefit the people of Nigeria, but rather a rudimentary system of administration that would facilitate the extraction of surplus from local producers. As William Graf has rightly observed, "Each region, according to its natural factor endowments and convertibility to colonial purposes, produced crops or minerals of a greater or lesser exploitative value."[19]

The people of the Northern Region were "encouraged" to produce groundnuts, cotton, and tin; the East produced palm oil, and the West cocoa. These were exported raw to Britain to be processed and turned into finished products, which were in turn brought back to Nigeria to be sold to the local people at exorbitant cost. Nigerians, right from the onset, were coerced through a variety of colonial policies to produce what they did not consume and consume what they did not produce. This fundamental disarticulation in the national economy was to be exacerbated with the advent of the "oil boom" in the seventies, and subsequently triggered the economic crisis of the early eighties, which gave the IMF and other Western multilateral financial agencies the opportunity to impose the Structural Adjustment Program (read debt peonage) on Nigeria in 1986.

The major feature of Nigeria's political economy from the 1970s to date is oil, its production and marketing. It has been estimated that the country earned $101 billion in oil revenue between 1958, when production commenced, and 1983.[20] This is no mean sum. The country's postcolonial elite, who had taken over from the departing British through a gentleman's agreement whereby the inherited dependent, peripheral-capitalist structure of the economy was left intact, was unable to transform itself into a truly self-reliant national bourgeoisie, however, and play its historical role—like their European counterparts in the nineteenth century—and create a truly national economy. Rather, this successor elite chose to become commission agents of the big commercial houses and mining companies that the departing British still controlled, while also moving to capture political power in order to use it as an instrument to secure more economic benefits for themselves.

Thus, the enormous oil revenue was not only squandered by this elite class that lacked any historical raison d'être; the agricultural sector—which between 1960 and 1974 contributed more than 50 percent of the revenue of the three defunct regions—was neglected. There is a positive correlation between the sharp rise in oil revenues beginning from the early seventies and the slump in agricultural production.[21] Indeed, between 1970 and 1982, yearly production of the main Nigerian cash crops—cocoa (the world leader in 1960), rubber, cotton, and groundnuts—fell by 43, 29, 65, and 64 percent respectively. The country's rate of food production also dropped between 1975 to 1983. But the elite were not merely content with cornering the oil revenue. Like their colonial predecessors, they also extracted the surplus of the peasantry through the agency of state-controlled marketing boards, which bought the latter's produce cheaply, exported it to Europe, and diverted the proceeds to private bank accounts. Investments were not made in the key Renewable Natural Resources sector of the economy—agriculture, forestry, and fisheries. The government began to rely more and more on oil revenue and foreign borrowing to finance its imports. The Dutch Disease—in Nigeria's case the sharp decline in the tradable agricultural sector—had set in.[22]

Terisa Turner has applied Ruth First's concept of the "Rentier State" to Nigeria's post-civil-war political economy, pointing out that the country is sustained not by what it produces, but on "rent" on production: here, the oil industry, where investments, production, marketing, and sundry expertise are completely dominated by multinational corporations that simply pay taxes and royalties to the state. Thus the entire state apparatus becomes a commodity for rent to the highest bidder, a bizarre bazaar presided over by a "commercial triangle" of state officials, local middlemen, and foreign suppliers.[23] This group cannot thrive outside the state political economy, which partly explains the dizzying succession of military coups and electoral frauds the country has been afflicted with since independence. State power is everything, and to be without power is to be condemned to unremitting poverty.

General Gowon's military governors made looting the national treasury their favorite pastime, while the Military High Command in turn awarded soldiers salaries eight times more than the national average.[24] The military government's halfhearted attempt to indigenize important sectors of the economy in 1972 and again in 1977 was, as Claude Ake has explained, subverted by a clique of senior military officers, "super" permanent secretaries, and commission agents who acted as fronts for the multinational companies whose stranglehold on the national economy the Nigerian Enterprises Promotion Decree of 1977 was supposed to break.[25] By the time the soldiers handed power back to the civilian government of Shehu Shagari in October 1979, the fundamental problems of Nigeria's political economy—disarticulation, dependence on foreign imports including food, and inequality, were yet to be tackled. Nigeria was effectively an enclave economy, relying on an oil industry that generated few jobs and had virtually no structural self-sustaining linkages with the other sectors of the economy, which in any case had lapsed into the doldrums. Gavin Williams graphically described the disease that afflicted the country's political economy in the oil boom years:

> The state has promoted the development of capitalism, foreign and domestic, by shifting resources from more competitive to less competitive producers, from craft to factory production, from agriculture to

> industry, from rural to urban areas, from the poor to the rich, and from Nigerians to foreigners. It has hardly given free rein to the ability of the people to produce goods. It has promoted the "wealth of the nation" but only by the impoverishment of the people.[26]

The politicians of the Second Republic did not fare any better than the soldiers they took over from in October 1979. Although oil export revenues reached an all-time high of $24.9 billion in 1980, it was clear that the economy, dependent on oil receipts for 96 percent of its external revenue—even though the sector accounted for a mere 27 percent of GDP—and also saddled with an external indebtedness that had ballooned to $9 billion, was heading for trouble. The oil boom was effectively over. The Shagari government's attempt to rein in imports and restructure the economy along lines of self-reliance with a series of "austerity" measures in 1981 was largely ignored by a political class that was busy looting the treasury. Although oil revenues between 1979, when President Shagari took office, and December 1983, when he was ousted, were about $800 million more than what was realized between 1960 and 1979, his government still managed to leave behind a foreign debt of $16 billion.[27] Tom Forrest has pointed out that capital flight from Nigeria first peaked during Shagari's tenure.[28] Everybody wanted a piece of the action, not least British Aerospace, which concluded a controversial £22 million ($35 million) kickback deal with Nigerian government officials in order to secure a contract for the supply of eighteen Jaguar ground-attack fighters worth £300 million ($480 million).[29] It was in fact the "Jaguar deal" that finally pushed junior military officers to demand an end to the Second Republic and subsequently brought General Muhammadu Buhari to power after a military coup in December 1983.

The economy that Buhari took over was leaking like a sieve. Twenty percent of the nation's oil was being smuggled out of the country. Western creditors were baying at the gate and, through the IMF, were exerting pressure on the government to ram through a package of economic measures that would enable the country to pay back what they claimed it owed them. However, as the Johnson Mathey Bank affair was later to reveal, a substantial portion of these "debts" was bogus, as British banks, Asian merchants, and Nigerian officials had colluded to defraud the country of billions of pounds of oil revenue.[30] An increasingly desperate Buhari appealed to Margaret Thatcher's government to help his regime recover the equivalent of over $8 billion that corrupt Nigerian government officials had salted away in British banks. Incidentally, this was the amount Western banks claimed Nigeria owed them. But Thatcher replied by threatening to publish the names of all Nigerians who had accounts in U.K. banks. The new military government quietly dropped the request.[31]

By mid-1985, 44 percent of Nigeria's entire export earnings was going into the servicing of these debts. Still, the Western creditors were not satisfied. They wanted the government to take an IMF bridging loan that amounted to almost $5 billion, along with a set of harsh conditionalities that would throw Nigeria's monocultural economy even more wide open to Western imports and further impoverish the people. When Buhari resisted this and instead embarked on a countertrade program, exchanging Nigerian oil for vital imports from selected countries in a bid to escape the financial stranglehold the Western creditor-nations had imposed on the country, a coalition of local and international oil and banking interests eased him out of power and, in August 1985, put General Ibrahim Babangida, a trusted ally, in his place.

Rounding the Circle

If Lugard and Goldie Taubman represented the crude imperialist stage of Western capitalism in Nigeria, the urbane and politically astute General Babangida represented its sophisticated but infinitely more vicious neocolonial phase. The Structural Adjustment Program that Babangida introduced in October 1986 devalued the naira, removed subsidies on key social services, and also set in motion the privatization of government-owned companies and agencies. SAP as a policy instrument to deregulate the economy of an underdeveloped country and make it more efficient was poorly thought out and even more shoddily implemented. Besides, the regime and its leader were noted for corruption and financial indiscipline, and it was only a matter of time before stagflation set in and the bottom dropped out of the national economy. Companies and other employers of labor, caught in a vicious cycle of low productivity and rising costs, began to retrench drastically. Food and other consumer goods disappeared from the markets and shops. When they were available, they were simply out of reach of the urban poor and millions in the rural areas who bore the brunt of Babangida's harebrained economic policies. Starvation and disease swept through the urban ghettos and the countryside, reaping a grim harvest. The nation had never had it so bad.

By the beginning of 1989, four years into Babangida's dictatorship, anger and resentment had begun to rise. There were sporadic street riots and work stoppages. In September, crowds led by university students poured out into the streets to protest against the government's structural adjustment policies and the mass suffering and poverty it had engendered. These demonstrations were brutally suppressed by an obdurate military junta that had long ago lost touch with the mass of the people. Meanwhile, the oil-producing communities of the Niger Delta, long oppressed and denied the oil revenue that was rightfully theirs, were smoldering. Already impoverished and deprived of the basic necessities of life by successive governments, they felt the adverse effects of Babangida's disastrous policies particularly keenly. The people of Iko, an oil-producing community, had organized a peaceful demonstration in 1987 to protest the exploitation of their oil by Shell and the federal government. This was brutally suppressed. In October 1990 the inhabitants of Umuechem, another oil-producing community, took to the streets, and armed troops, called in after Shell's request for "security protection," killed and maimed them into submission. Then the Ogoni launched the Movement for the Survival of the Ogoni People (MOSOP), to put a stop to this regime of ecological devastation and economic exploitation. Finally, the chicken was coming home to roost.

OMPADEC and the Cult of Corruption

Pressed by Shell officials who correctly divined that a storm was brewing in the oil-producing communities, General Babangida responded to the Umuechem massacre and other "disturbances" in the Niger Delta with the Oil Mineral Producing Areas Development Commission (OMPADEC), established by Decree 23 of 1992. The decree increased the 1.5 percent of oil proceeds allocated to the oil-producing communities to 3 percent, and transferred the fund to the new commission to administer on their behalf.[32] OMPADEC was conceived as a development agency for the oil-producing areas and was charged with the responsibility of monitoring and managing ecological problems associated with the production and exploration activities of the oil companies. It was also expected to act as mediator between them and the communities when problems arose.

While these are obviously laudable goals, the manner through which the commission came into existence and the hidden motives of its promoters were such as to stymie from the outset its evolution into a genuine development agency for the oil-producing communities. The government was determined to ensure that the commission was just another lame public agency and that the actual funds that would eventually reach the communities would be just enough to keep them quiet.[33] It is instructive that Albert K. Horsfall, appointed executive chairman of OMPADEC in July 1993, was a senior operative of the State Security Service (the notorious SSS) before he was "redeployed" to establish the commission's headquarters in Port Harcourt. Horsfall was to report directly to the Head of State, General Babangida. Although OMPADEC had a budget of approximately $95 million in 1993—at least on paper—money was disbursed to it at the Head of State's whim and so haphazardly as to make proper project planning simply impossible. A government report stated that for his part, Horsfall operated as a law unto himself, and considered the oil-producing communities as a private fief.

Since OMPADEC was under no supervisory authority other than the presidency, which in any case had "weightier" matters of state to engage its attention, inefficiency and financial mismanagement quickly set in. The World Bank team that studied OMPADEC's activities in the Niger Delta communities in 1995 concluded that it would be difficult for the commission to effectively fulfill its role as a development agency because (1) there was no emphasis on environmentally sustainable development; (2) the commission did not have the requisite personnel to enable it to meet its ecological mandate; (3) there was an absence of long-term planning; (4) there was little or no project assessment, and where projects were initiated, maintenance requirements were not built into them; and (5) there was no integrated approach to development planning, which should have involved the local communities and other government agencies in the area.[34]

Several community leaders in the Niger Delta also dismissed OMPADEC as another white elephant project established by the government to lure them into a false sense of contentment, believing that at long last something was being done to redress past wrongs. Claude Ake, a development expert, criticized OMPADEC for being too overcentralized and predicted that nothing good would come out of it. Ken Saro-Wiwa also argued that the commission had come too late in the day, and wanted to know how OMPADEC intended to address the case of such areas as Ogoni, whose oil fields originally produced 300,000 barrels daily but were now down to 30,000. "Three percent of what are we getting?" Saro-Wiwa asked. "What happens when the oil dries up?"[35]

But perhaps what proved OMPADEC's greatest shortcoming was the use it was put to by Shell and government officials alike. The latter saw OMPADEC as yet another avenue for corrupt self-enrichment. Indeed, as soon as it was established in 1993, OMPADEC quickly developed a reputation as a government agency where favors were sought and sold for cash. Even Shell officials, who are probably no saints where such matters are concerned, complained that the money given to OMPADEC was not properly accounted for. A senior Shell official who spoke to Human Rights Watch investigators in Lagos in March 1995 said, "We pay a lot of money to OMPADEC and there is no return."[36]

Following loud complaints from leaders of the oil-producing communities, who accused OMPADEC officials of "neglect, ineptitude, insensitivity, high-handedness, corrupt practices, and autocratic style of leadership," the federal government set up a four-man investigative panel headed by Eric Opia, a politician and confidant of General Sani Abacha, who took over as chairman of the commission after Horsfall was sacked in February 1996. The panel, charged with inspecting, assessing, and reporting on the projects and finances of the commission, undertook a tour of the various projects. Its report, when it was finally submitted to the government in December 1996, opened a veritable can of worms. Said Mike Akpan, a journalist who studied the voluminous report and wrote a magazine article based on its findings: "Virtually all the various interest groups complained that projects which did not necessarily represent what they wanted most were imposed on them and that even most of the projects were left uncompleted." He added that the panel's "findings during the tour were not only revealing but also mind-boggling. Indeed it was a reflection of financial recklessness in the award of contracts."[37]

Abandoned and poorly implemented development projects presently dot the Niger Delta, courtesy of OMPADEC. One example is the Eleme Gas Turbine Project in Port Harcourt, which was awarded to Marshland and Project Nigeria Limited, an engineering company, in June 1993 for the contract sum of $20.7 million. As originally conceived, the gas turbine project was designed to supply electricity to all parts of Rivers State, including some thirty communities in the three local government areas of Ogoni. However, four years after the plant was "commissioned" by the Head of State in October 1995, the Eleme Gas Turbine had yet to generate any electricity.

One of the first contracts Albert Horsfall awarded when he assumed office as chairman of OMPADEC in June 1993 was the Port Harcourt Water Project, contracted out to ICER Nigeria Limited. The total contract sum was for $4 million. Curiously, though, the contractor was paid less than $3 million—representing 70 percent of the total contract sum—even before the project was started, ostensibly to enable ICER to mobilize men and material for the work. Following the sacking of Albert Horsfall, who it was discovered had connived with the company to divert OMPADEC funds to other uses, ICER was taken to court by officials of the commission. Fearing exposure, the company's officials quickly indicated their willingness to engage in dialogue and settle out of court. However, it is unlikely that OMPADEC will recover the money that Horsfall paid out to ICER.[38] By the first quarter of 1996, three years after it commenced operations, OMPADEC had committed itself to projects worth $530 million. Interestingly, the bulk of money paid out for projects "completed" was to contractors whose addresses could not be traced. When Eric Opia, head of the panel set up by the Abacha junta to probe Horsfall, was appointed Sole Administrator in his place, he proceeded to loot OMPADEC in an even more brazen fashion. By September 1998, when he was kicked out for "gross financial misappropriation," Opia had embezzled some $200 million set aside for the development of the impoverished communities of the Niger Delta.

For its part, Shell's attitude toward and relationship with OMPADEC was cynical and mercenary, using the commission as yet another instrument to dominate the oil-producing communities. The collusion between the two organizations became public knowledge in the course of Ken Saro-Wiwa's trial in 1995 when two prosecution witnesses confessed that officials of Shell and OMPADEC in league with others had bribed them to testify against the MOSOP leader and implicate him in the murder of the four Ogoni chiefs in May 1994.[39] A leaked memo originating from Lieutenant Colonel Paul Okuntimo, the former commander of the Rivers State Internal Security Task Force—specially created in January 1994 to identify, isolate, and eliminate leaders of MOSOP and other activists in the Niger Delta leading the resurgent movement for social and ecological justice in the area—also seemed to implicate Shell, other oil companies, and OMPADEC in a proposed military operation targeted at the oil-producing communities.[40] Shell has consistently denied that its officials bribed prosecution witnesses during the trial, or that they worked at any time with Okuntimo's security task force to repress MOSOP.

A Game for "Presidents"

Oil and its by-products have always provided Nigeria's military leadership a lucrative source of unearned income. While corruption and the misappropriation of the country's huge oil revenue is the favorite pastime of the military and civilian elite, General Ibrahim Babangida, who gave himself the title of "President" as soon as he seized power in August 1985, turned corruption into an industry, and in the process cornered billions of dollars of the oil revenue for himself. The Persian Gulf Oil crisis of 1990–91 was exactly what Babangida needed to make his final killing. The Gulf War triggered a sudden and sharp rise in international oil prices, but instead of spending the additional revenue in productive social and economic projects, Babangida and his cronies saw the windfall as a personal bonus. According to Kayode Fayemi, "The extra fund was regarded as discretionary income which went on a massive spending binge that diverted revenues into corruption-funded patronage, sharply expanded extra-budgetary expenditure, and bloated an already inflation-ridden economy."[41]

Shortly after General Babangida was ousted from power in August 1993, his successor, General Sani Abacha, set up a panel headed by the respected economist Dr. Pius Okigbo to look into the finances of the Central Bank of Nigeria during the Babangida years. In his statement on the occasion of the submission of the report of the Panel on the Reform and Reorganization of the Central Bank of Nigeria in July 1994, Dr. Okigbo accused Babangida and members of his government of stealing the country blind:

> Between September 1988 and 30 June 1994, US$12.2 billion of the $12.4 billion [in the dedicated accounts] was liquidated in less than six years . . . they were spent on what could neither be adjudged genuine

> high priority nor truly regenerative investment; neither the President nor the Central Bank Governor accounted to anyone for these massive extra-budgetary expenditures . . . these disbursements were clandestinely undertaken while the country was openly reeling with a crushing external debt overhang.[42]

William Keeling, the Lagos correspondent of the *Financial Times* who investigated the Gulf Oil windfall scam and published the story in 1991, was set upon by state security operatives and deported on Babangida's orders.[43] The military junta also promulgated a decree to stamp the deportation as a permanent legal fixture in Nigerian legal jurisprudence.

Following growing pressure by popular democratic forces for Babangida to conduct elections and hand over power to elected representatives of the people, the military president finally scheduled presidential elections for June 12, 1993, after a long-winded political transition process that saw several postponements, detours, and the mass banning of credible members of the political class. At the last moment, though, the sheer thought of giving up power and, consequently, the billions of dollars of oil loot that went to his private bank accounts, proved too much for Babangida. Aided by ambitious Young Turks in the Army, he annulled the results of the presidential elections which his friend and business associate M.K.O. Abiola had won. It was, however, too late to stem the pro-democracy tide, and Babangida was forced to relinquish power after putting in place an interim national government headed by Ernest Shonekan. The only legacy that Ibrahim Babangida bequeathed to Nigerians before he was removed from office was the democratization of corruption and the corruption of democracy.[44]

Military coups in Nigeria are a zero-sum game. If you succeed, the prize is instant access to the billions of dollars of oil revenue extracted from the Niger Delta annually. If the coup is botched and you are caught alive to boot, the penalty is a swift court-martial and summary execution. The glittering oil prize has, however, always proved an irresistible pull for Nigeria's ambitious and largely indolent officer corps, who, given the chance, are willing to walk in the Valley of Death to seize it. And so on November 17, 1993, General Sani Abacha, who was Babangida's second-in-command and had participated in two previous coups, staged his own, sweeping the interim government contraption aside and assuming full powers as military Head of State. An infantry soldier who rose through the ranks to his present position, Sani Abacha was feared by allies (he had few friends) and foes alike for his ruthlessness and animal cunning. He personally ordered troops into the streets of Lagos during the mass demonstrations to protest the annulment of the June 12 presidential elections. By the time the sound of gunshots finally faded away, hundreds of unarmed students, youths, and children lay dead or dying. When the country's oil workers went on strike in August 1994 to press for the restoration of the election results, General Abacha crushed the protests with uncommon but characteristic brutality.

Like his predecessor, General Abacha also made his own pile in the course of his long and checkered career as a military officer adept at coup-plotting. He was widely perceived to be corrupt, ruthless, and narrow in his world-view.[45] Again like his former boss, he proceeded to divert the oil wealth of the Niger Delta to his own private ends. Nor was he sympathetic to the problems of the oil-producing communities, as the military crackdown that commenced in the area as soon as he seized power amply demonstrated. Not only did Abacha make the Niger Delta-based OMPADEC redundant by freezing its bank accounts—ostensibly to inspect the commission's books—he set up another development agency, Petroleum (Special) Task Force, also funded from oil proceeds, and appointed General Muhammadu Buhari, a former Head of State, its chairman.[46] Again like OMPADEC, certain officials of the PTF, established by decree and inaugurated in March 1995, have been accused of nepotism and financial recklessness.[47]

The International Monetary Fund (IMF) had reported that some 150,000 barrels of oil produced daily in Nigeria was not accounted for, implying that the proceeds had been diverted by the general and his henchmen. In September 1995, Abacha ordered the revision of allocations of crude term contracts, giving the controversial Swiss-based oil trader Marc Rich and his front company in Nigeria, Glencore, control of a third of the country's term crude supplies. Glencore, before its grip on the Nigerian oil industry was loosened in June 1999, held three of the five lucrative crude-for-products countertrade deals, operated one condensate contract, and had four of the six oil import contracts for an estimated 55,000 tons a month.[48] Marc Rich, interestingly, was General Abacha's business partner, and they shared a mutual associate in Gilbert Chagouri, front for the general's extensive business empire.[49]

General Abacha was on the verge of transforming into a "democratically elected" president when he died on June 8, 1998, apparently of heart failure, in the arms of two prostitutes in the presidential mansion in Abuja.[50] He had, after several postponements, promised to hand over power to a new "democratic" government in October 1998. He decreed five political parties into existence, all led by front men. Prior to this, however, he had unleashed state security operatives on democracy and human rights activists, journalists, and politicians who dared speak out against his excesses. He had put Chief Abiola, winner of the June 12, 1993, presidential election, in jail after the former declared himself president in June 1994. Several of Abiola's supporters, including his senior wife, Kudirat Abiola, were murdered by Abacha's special hit squad.[51] Detention centers, some of them equipped with special torture chambers designed to inflict the maximum pain and extract "confessions" from "subversive elements," sprang up all over the country. Abacha's security operatives rapidly filled them with radical journalists, trade union leaders, and democracy activists. By the time of the dictator's death in June 1998, there were over eight thousand political prisoners in his gulag, including General Olusegun Obasanjo, a former military Head of State, and his deputy, General Shehu Yar'Adua. The latter was to die in detention, allegedly poisoned by state security officials acting on Abacha's instructions.

General Babangida had looted the treasury using a sophisticated array of fronts and devices to cover his tracks. Abacha dispensed with such niceties. His first priority, as soon as he seized power in November 1993, was to bring the Central Bank of Nigeria and the Nigerian National Petroleum Company, the two vital organs through which the country's estimated yearly oil revenue of $8 billion was processed, firmly under his control. Gilbert Chagouri, eldest son of a Lebanese family with extensive business interests in Nigeria, was the conduit through which Abacha siphoned billions of dollars abroad. It has been estimated that $10 billion was looted by Abacha and his cronies between December 1993 and June 1998.[52] In November 1998, General Abacha's successor, General Abdulsalami Abubakar, announced that his government had retrieved $1 billion—10 percent of the loot—from Abacha's family. An additional $250 million was recovered from Ismaila Gwarzo, the late dictator's national security adviser.[53]

General Abubakar was no better than his predecessor. Although he freed the bulk of the political detainees, he refused all entreaties to set Chief Abiola, the country's elected president, free. On July 7, 1998, one month after

Abacha's demise, Abiola died in suspicious circumstances while still Abubakar's prisoner.[54] Following Chief Abiola's death, the new military strongman announced a new political transition program to end in May 1999, with the transfer of power to a democratically elected government. Three political parties—the People's Democratic Party (PDP), All Peoples' Party (APP), and the Alliance for Democracy (AD)—were registered to contest state and national elections. On May 29, 1999, Olusegun Obasanjo, a former military Head of State, was sworn in as president after a controversial election that was attended by accusations and counteraccusations of vote-rigging, bribery, and thuggery—most notably in the Niger Delta area, where ballot boxes were stuffed with fake ballot paper by local politicos at the behest of officials of Obasanjo's party, the People's Democratic Party.[55]

President Obasanjo inherited an empty treasury, however. While the politicians were traversing the length and breadth of the country in search of votes, General Abubakar and his generals indulged in a last-minute orgy of plundering the treasury. First, Abubakar awarded to himself, the disgraced former Head of State General Ibrahim Babangida, and a handful of senior generals and business associates—including the ubiquitous Gilbert Chagouri—eleven oil exploration blocks and eight oil-lifting contracts worth billions of dollars.[56] Then the generals turned their attention to the country's foreign reserves. In the short space of three months—between the end of December 1998 and the end of March 1999—$2.7 billion had vanished from the national coffers. When the influential London-based newsletter *African Confidential* blew the whistle on this financial hemorrhage in mid-April, Abubakar's finance minister tried to explain that the economy had faced extraordinary financial demands, including the drastic fall in the oil price, the cost of the elections and return to civilian rule, as well as the cost of Nigeria's military leadership of Ecomog, the West African peacekeeping force in Sierra Leone. However, as *African Confidential* has correctly pointed out, "The extra cost of the Ecomog operations was not foreseen in the budget calculations, but would not account for much of the $2.7 billion drawdown. Nigeria's operations in Sierra Leone are said to cost $1 million a day. Even if that is doubled after the rebel advance on Freetown, it would have added just $90 million to the extra-budgetary expenditure, already partly defrayed by contributions from Britain and the United States. As for payments to service foreign debts, they have fallen well short of the level set in the 1999 budget."[57] *Africa Confidential* concluded that

General Abubakar, along with a handful of senior army officers and civil servants, may have dissipated a staggering $2.6 billion in covert foreign exchange deals—money that ended up in their private bank accounts. This has been confirmed by Chris McGreal, the Africa correspondent of *The Guardian of London*.[58]

President Obasanjo has promised "decisive action" to deal with the country's financial collapse. Although he has set up a panel to review the slew of contracts General Abubakar awarded to himself and his cronies in dubious circumstances in the last days of his regime, and also pledged to introduce a wide-ranging anticorruption bill, Obasanjo has so far shown no great enthusiasm in taking action where it really matters: getting former military officers to account for their financial misdeeds. Observed Chris McGreal, "[Obasanjo] has declined to target the officers who plundered billions while in power, among whom the most notable is General Ibrahim Babangida, a former military ruler who has yet to explain what happened to a windfall of nearly $12 billion that came Nigeria's way when oil prices surged during the Gulf war."[59] Obasanjo did, however, dissolve the boards of OMPADEC and PTF, preparatory to establishing a new Niger Delta Development Commission in their place.

One of the first things Obasanjo did on assuming office was to visit the oil-producing communities of the Niger Delta, in the first week of June 1999, shortly after government officials announced that the new constitution had increased to 13 percent of accruable federal oil revenue the amount of money to be spent in developing the area. His political party, however, has no clear policy position on how to tackle the horrendous poverty and environmental devastation visited on the people these past four decades by Royal Dutch Shell and successive Nigerian governments. If the representatives of the local communities who met with Obasanjo in Port Harcourt expected a concrete program of action from the new civilian ruler, they were disappointed. Said Felix Tuodolor, president of the Ijaw Youth Council, an organization fighting peacefully to achieve self-determination and environmental and social justice for the Niger Delta communities, "Our people have long passed the stage of niceties and empty words. We will begin to take Obasanjo seriously only when he has told us what happened to the billions of dollars of oil revenue that were taken away from the Niger Delta. Thirteen percent is not the answer. We will take him seriously only when he withdraws the soldiers that even now are killing and maiming our young men and raping our women and puts in place a new political arrangement that satisfies our demand for self-determination. But we are tired of waiting. Our land has been plundered and ravaged. Our rivers and creeks are dying. It would be criminal to expect us to fold our arms and wait for the oil companies and their allies in government to deliver the last death-blow. We won't."[60]

THREE

Colossus on the Niger

Since the beginning of Shell's operations in the Niger Delta, the company has wreaked havoc on neighboring communities and their environment. Many of its operations and materials are outdated, in poor condition, and would be illegal in other parts of the world.

Greenpeace[1]

"The Most Profitable Company in the World"[2]

Royal Dutch Shell began life in 1907 following the merger of the British firm Shell Transport and Trading Co. (STTC) and Royal Dutch Petroleum Co. of Holland. The company has grown constantly, diversifying its holdings, stretching its tentacles to virtually all countries of the world, and decentralizing its operations to such an extent that even some of the company's senior managers run into difficulty trying to explain precisely how Royal Dutch Shell functions.[3]

Every year Shell competes with Exxon for the title of the world's biggest oil company. The company, however, has a greater geographical spread than its American rival. In 1979, Shell became the first company in the world to post sales of $3.2 billion over a twelve-month period.[4] By 1996 the Shell Group's annual profit had leaped to $9.1 billion from sales of $176 billion. Said Mark Moody-Stuart, chairman of STTC, on the occasion of his company's centenary celebrations in November 1997 in London, which the Queen attended, "Were our founder, Marcus Samuel, to reappear today, I do not think he would be displeased with what has grown from his efforts."[5] Shell produces oil and gas in forty-five countries, but it also has interests spread around a hundred countries, in biogenetic engineering, petroleum marketing, and the mining of coal, bauxite, zinc, uranium, aluminium, copper, nickel, gold, and lead.

The structure of the conglomerate reflects the complexity and the diversity of its interests and areas of operation. The two parent companies, Royal Dutch Petroleum Company of Holland and Shell Transport and Trading Company, registered in England, own 60 percent and 40 percent respectively of two holding companies, Shell Petroleum NV and Shell Petroleum Co. Ltd. These holding companies provide the umbrella for the conglomerate's several service and operating companies, among them Shell International Petroleum Maatschappij BV; Shell International Chemie Maatschappij BV; Shell International Petroleum Company Ltd.; Shell International Chemical Company Ltd.; Billiton International Metals BV; Shell International Research Maatschappij BV; Shell International Gas Ltd.; Shell Coal International Ltd.; and Shell International Marine Ltd. The bulk of these companies are 100 percent owned by Royal Dutch Shell. Queen Beatrix of the Netherlands and the Queen of England are major shareholders.[6]

Gulliver on the Rampage

Shell employs a sophisticated array of damage-control experts, scenario planners, lobbyists, and spin doctors to present the image of a caring, thoughtful, and socially responsible company to the outside world. Long before the issue of the environment became a topic of national discourse in Europe and the United States, and multinational oil firms were forced to adopt the veneer of environmentally friendly companies, Shell had elevated the concept of selling itself to the powerful conservationist lobby into an art form, devoting a considerable chunk of its budget to this effort over the years.[7] So successful was this portrayal, and so skillful were Shell's image makers in contriving a semblance of "constructive dialogue" and openness toward critics, that it took a major public relations disaster like the company's intention to drill for oil in a protected forest in rural England in 1987 for this carefully cultivated image to shatter and for conservationists and ecologists to see Shell for what it really is: a modern-day

Gulliver on the rampage, waging an ecological war wherever it sets down its oil rig.[8]

Shell's acquisition, and subsequent despoliation, of the land of helpless people all over the world actually commenced after the company began oil production in British Borneo, Mexico, and Venezuela before the First World War. Shell literally bulldozed its way into the lands of the Quicha, Achual, and Shuar people in the Ecuadorian Amazon in the 1920s.[9] The company constructed a road running through the area in the course of oil exploration, opening it up for subsequent invasion by European colonists and missionaries who in turn pushed the Indians out of their lands and imposed debt peonage on them through unfair trade practices.[10] In an attempt to fight off Shell and the fortune hunters that followed in its wake, the Shuar Indians formed a federation in 1964 to enable them to present a united front against the juggernaut in their midst, and today still see Shell, along with Texaco, as a major invader of their territory.[11]

The conglomerate made history in 1978 when it became the first multinational mining company in the world whose headquarters was visited by an official delegation of aggrieved aboriginal peoples.[12] Billiton, a Shell subsidiary, had acquired a bauxite mining lease of 736 square miles in North Queensland in Australia in 1975, along with two American mining companies. Two years later, despite the opposition of the Aurukun people who owned the land, the Queensland government gave Billiton and its associates rights over Aurukun hunting and pastoral land and also granted them permission to build a company town and a harbor, as well as rights to mine the surrounding coastal reefs, including the mainland. The lease was to last till the year 2038. The people of the Aurukun Aboriginal Reserve, however, decried the company's attempt to mine bauxite on their land without their consent, arguing that the presence of the mines would eventually lead to the pollution and subsequent extinction of Aurukun culture and way of life. They took their case to the headquarters of Billiton International at The Hague in November 1978, and following a major campaign in the Dutch media, the company appeared to back down, promising that it would not venture into the Aurukun reserve without consulting with the people.

This was, however, a tactical retreat. As Bernard Wheelahu, Billiton's Metals Manager in Australia, was to declare two years later, "I prefer not to give land rights to the Aborigines . . . it means they will be in the same position as the other white Australians [*sic*] . . . It is a very big problem and it is dangerous to the mining industries."[13] In all, the bauxite deposits were worth a conservative $27 billion (in American dollars), but all the Aurukun got was the equivalent of a derisory eight dollars per square mile as rent (with the promise that this would rise to about fifty dollars after fifteen years) and a royalty of one dollar a ton. The Aborigines did not find this acceptable, and they took their case to the courts, demanding a direct share of the royalties, guarantees of employment if the bauxite project proceeded, guarantees that their sacred places would not be defiled, and equality of treatment with the whites. They eventually lost their case in the London Privy Council and the Queensland government handed over the lands of the Aurukun to Shell/Billiton to do with as it pleased.[14]

The indigenous peoples of South America have also received a good dose of Shell's methods. Manu National Park in Peru is internationally recognized as one of the largest ecoresource areas in South America. It is also home to several indigenous peoples.[15] In 1981, Shell, in collaboration with GeoSource Inc., entered the Fitzcarrald Isthmus, which is located in the park, and began seismic exploration for oil, ignoring the protests of local inhabitants and conservationists alike. The indigenous peoples of the area, particularly the Nahua, have been devastated by the oil company's activities on their land. In the course of preliminary oil exploration in Nahua territory in the mid-eighties, Shell opened up the frontier rain forest, causing the Nahua to be exposed to outsiders who brought with them diseases that the former could not cope with. According to Project Underground, a San Francisco-based NGO, it has been estimated that between 30 and 50 percent of the exposed population was wiped out as a result of these diseases. Shell denies responsibility for this tragedy but has not produced independently verifiable evidence to support its contention.[16]

In 1997 the company began oil production at its Cashiriari 2 site, one of ten planned oil exploration campaigns located on indigenous peoples' legally recognized territories—including the Nahua and the Kugapakori. Although Shell conducted an environmental impact assessment and developed an environmental management plan (EMP) for its operations, the local communities are already crying out that fuel spillage occurs regularly at Shell's logistics operations center in Nuevo Mundo. It was also reported that at least five of these spills entered the Urubamba River in February 1997 alone.[17]

The Shell environmental chief in the area, Murray Jones, and terrestrial ecologist Miguel Ruiz-Larrea admitted to journalists and environmental activists that the company's activities had indeed adversely impacted on Nuevo Mundo village and environs, but argued that population pressure should also be factored in the process of environmental change in the area. Said Ruiz-Larrea, "I don't want to say that our operations are not having an impact, because it is clear they are, but I think there are a number of other factors involved." Even more significant, Shell has put in place a new health program in its Cashiriari project to ensure that all its workers and visitors are vaccinated so that new diseases do not enter the region, leading to a repeat of the Nahua tragedy. But Peruvian activists like Doris Bavin, an environmental lawyer, have complained that Shell has overlooked its moral responsibilities to the local people, and that they do not have an idea of what is happening at the drilling site or what the future operations might involve. Project Underground also commended Shell's commitment that Peru's indigenous populations would not be harmed in the course of the new project, but added that "they are meaningless if they are not enforced. We have real reasons to doubt Shell's sincerity."[18]

Elsewhere, in Brazil, where it is the largest privately owned company in the country, Shell has penetrated the Aripuana National Park in the Middle Amazon, which was originally reserved for Indians, shrugging off all opposition and setting down its rigs to drill for oil and gas. The Tucurui hydro scheme, a project on the Tocantin River in which Shell is involved, laid waste to vast swaths of the jungle. Toxic chemicals were used to clear the flora and the containers were carelessly discarded. They were later put to use by the local people—and this led to miscarriages, food poisoning, and deformities in children.[19]

In Bangladesh, where the military is waging a genocidal war on the tribespeople of the Chittagong Hill Tracts in a bid to drive them out of their mineral-rich land, Shell, working with the nationalized oil company Petrobangla, completed seismic surveying of the area in 1985 despite spirited campaigns mounted by the NGO Survival International to get the company to withdraw. The company has also, over the years, acquired or leased vast tracts of Native American land, totaling 258,754 acres in 1983 and containing over twenty billion tons of coal.

Apartheid's Handmaiden

Shell's role in aiding and sustaining the apartheid regime in South Africa has been adequately chronicled, beginning in 1979, when antiapartheid journalists and researchers in London and Amsterdam blew its elaborate cover and revealed to the world that the oil company had been violating the OPEC embargo all through the seventies by using various fronts, notably Japanese and Swiss companies, to supply crude to South Africa. It was also revealed that Shell was the major beneficiary of the incentives that the apartheid regime put in place to tempt oil companies to break the embargo and sell oil to the country. International opposition to Shell's involvement in apartheid South Africa peaked in 1986 following a series of boycott actions against Shell filling stations in Europe, the United States, and several other countries, culminating in the attack on sixteen Shell gas stations four days before the company's 1986 annual meeting, by Dutch antiapartheid activists.[20]

Shell's business interests in the country date back to the 1920s. The company has always maintained a cozy relationship with Afrikaner merchants in South Africa. Today the conglomerate owns 50 percent of the Sapref refinery, the country's largest, in Durban. It also has a 50-percent stake in Abecol, an asphalt manufacturing firm, and another 50-percent holding in the controversial Rietspruit open-cast coal mine in the Transvaal. Shell also owns over eight hundred gas stations in the country.

Eighteen years after its link with the murderous apartheid state was exposed, the oil giant shows few genuine signs of making amends, coming to terms with demands for improved environmental performance, and behaving in a socially responsible manner. The Brent Spar incident, when the oil company attempted to dispose of an oil platform in the United Kingdom in 1995, not only shocked the world but also served as a grim reminder that Shell, despite its public campaigns to encourage the British people to improve the environment, does not always practice what it preaches.

Confidential Shell documents leaked to Greenpeace in Turkey in March 1996 revealed that contaminated waste from Shell's oil fields in Diyarbakir had been polluting underground drinking water.[21] The polluted aquifer is the only source of drinking water for over two million people in the area, and experts say the water could stay seriously polluted for three hundred years. Clearly, Gulliver is still on the rampage.

Colossus on the Niger

The corporate headquarters of Shell Petroleum Development Company of Nigeria (SPDC) is an imposing marble-and-glass skyscraper on the Lagos marina, Nigeria's bustling commercial capital. It is from here that the managing director of Shell Nigeria oversees the operations of the company, concentrated mainly in the Niger Delta.

Royal Dutch Shell has been in the country since 1937, when it began to explore for crude oil under the name Shell-D'Arcy, a joint venture between the oil conglomerate D'Arcy Exploration Company and the British colonial administration. Prospecting was stopped in 1941 because of the Second World War, to resume again five years later under the new name of Shell-BP Development Company. The company struck its first oil well in Oloibiri, an Ijaw village in the eastern part of the Niger Delta, in 1956. Commercial exploitation began two years later.

From this modest beginning Shell Petroleum Development Company of Nigeria is today the most important privately owned company in the country. Shell operates the largest oil-producing venture in the country in collaboration with the government-owned Nigerian National Petroleum Company and two other Western multinationals, Elf Nigeria, a subsidiary of the French oil company Elf, and Nigerian Agip Oil Company, a subsidiary of its Italian parent, Agip. Shell, the operating company of the joint venture, is solely responsible for day-to-day operations, and on its own, distinct from the three other venture partners, produces between 800,000 and one million barrels of crude a day, about half of Nigeria's entire daily oil production. To do this the company holds concessions over an area of 31,000 square kilometers in the Niger Delta, manages ninety-four producing oil fields and 3,800 miles of pipeline, and employs about 5,500 workers, including three hundred expatriates. A further 20,000 people work for the company either as subcontractors or temporary workers.[22]

Organizational Structure

The federal government of Nigeria, through the Nigeria National Petroleum Corporation (NNPC), owns 55 percent of the shares of Shell Petroleum Development Company (SPDC). Shell International Petroleum Maatschappij, a subsidiary of Royal Dutch Shell, owns 30 percent, while Elf, the French oil company, and Agip of Italy have 10 percent and 5 percent of the joint venture respectively. Since Shell is in charge of daily operations, SPDC enjoys the status of operating company alongside other Shell operating companies at the group's corporate headquarters in The Hague. Program discussions are held there at the beginning of every financial year, and SPDC, like the others, has to present and defend its investment plans and budget. The yearly program discussion is the pivot on which the entire operations of Royal Dutch Shell, its subsidiaries, and ancillary companies revolves. Indeed, the future of any operating company depends on how well it successfully defends its work plan for the year, as approval of its budget can be withheld by headquarters.[23]

In Nigeria, SPDC has its head office in Lagos, and the managing director is also the chief executive officer. The Lagos headquarters is, however, mainly concerned with administration, as the main areas of operations are located in the Niger Delta, in Port Harcourt, which is the operational base of the Eastern Division under a divisional manager designated General Manager East, and in Warri, base of the Western Division, led by the General Manager West. The activities of the two divisions are coordinated by the head office in Lagos, but the divisions also enjoy a measure of autonomy. There is also rivalry between the two divisions and they are encouraged to compete against one another, especially in the area of oil exploration and production targets.

A Veritable Money-Spinner

It is a measure of how important its Nigeria oil concessions are to Royal Dutch Shell that SPDC accounted for about half of the 93.1 million tons produced in the country in 1994—nearly 1.9 million barrels a day—at a market price of $16.20 per barrel. Between 1991 and 1995 the conglomerate's 30-percent share in SPDC generated between 250,000 and 290,000 barrels of crude a day, making Nigeria Royal Dutch Shell's third-biggest country of production after the United States and the United Kingdom. In 1994 alone, 11.7 percent of Shell's total crude oil production came from Nigeria.[24]

There is no doubt that Nigeria has been very profitable for Shell, but the precise figures in dollar terms are hard to come by. The company shrouds the financial side of its operations in Nigeria in mystery, using an elaborate cover of misleading statistics, vague statements, and sometimes outright hostility, to ward off prying eyes. However, going by figures that Shell itself supplied to journalists and environmental activists in London in 1995—which were by no means comprehensive—"Shell, Elf, and Agip share just one dollar per barrel."[25] According to Shell, a barrel of crude yields an average of $15, after the $2 used to produce it is deducted. Of this $15, the lion's share of $12 is taken by the Nigerian government, which owns 55 percent of the shares of the joint venture. The remaining $3, says Shell, is divided between investments in new oil exploration and production activities, tax on profit, which again goes to the Nigerian government, and the three other oil companies—Shell, Elf, and Agip—which amounts to "just" a dollar. Going by SPDC's share structure, two-thirds of the one dollar net profit per barrel for the three oil companies goes to Shell. Roughly translated in terms of the joint venture's oil production figures, Shell earns between $530,000 and $670,000 a day from its Nigerian concessions, amounting to an average of $200 million per year. Assuming we take these figures as a constant, and they are extremely conservative, the oil giant earned $2 billion between 1986 and 1995 from Nigeria.[26]

While this is a handsome profit by international standards, the above figures do not begin to reveal the full picture of Shell's earnings. Shell, along with the other oil companies, signed a Memorandum of Understanding (MOU) with the federal government of Nigeria in 1986 in an attempt by the latter to boost oil exploration and production in the country following the sharp fall in world oil prices. By the terms of the MOU, which took effect on January 1, 1986, the oil companies are guaranteed a profit margin of two dollars per barrel of oil they produce as part of the SPDC joint venture as long as prices oscillate between $12.50 and $23.50 a barrel. In 1991 the federal government increased this profit margin to $2.30 to $2.50 for companies that invest a minimum of $1.50 for every barrel of crude they produce. An additional ten to fifty cents has been set aside as further incentive to oil companies that provide evidence to show they discovered new oil wells with oil over and above the barrels they mined for a particular year.[27]

For Shell, therefore, the Nigeria oil industry is a winning game. This is so because international oil prices have for the most part stayed within the stipulated Nigerian government bracket since 1986, even rising above the $23.50 mark in some years. Together with the other foreign oil companies, the conglomerate has been earning $2.50 per barrel since the MOU was amended in 1991. Of this amount, Shell, Elf, and Agip plow back $1.50 toward further exploration and production. The other dollar is transferred to their corporate headquarters in The Hague, London, Paris, and Rome as profit for shareholders.

The tendency has been for Shell to present the money it reinvests in Nigeria for further oil exploration as cost rather than profit. This is misleading. As Greenpeace has pointed out, it is standard accounting practice to regard investment in new business activities (in the case of Shell, increasing its oil production) as profit. It has therefore been estimated, taking the MOU and Shell's additional earnings from it into consideration, that the company earns more than $500 million a year in Nigeria. Of this it reinvests $300 million in the country and transfers the rest to its head offices in The Hague and London.[28]

A new wholly owned Shell subsidiary, Shell Nigeria Exploration and Production Company (SNEPCO), was established in 1992. SNEPCO is presently exploring for oil off the Nigerian coast and in the Gongola Basin in the northern part of the country.[29] Shell is also the operating company of the new $2.4 billion liquefied natural gas project currently being developed by Nigeria Liquefied Natural Gas Limited near Bonny, and holds a 24 percent stake in it. Shell also has a 40 percent stake in National Oil and Chemical Marketing (NOCM), one of the leading marketers of petroleum and allied products in the country.

In 1992, Nigeria's proven oil reserves were projected at 17.9 billion barrels.[30] At the present rate of production it has been estimated that this will last another twenty-six years. But there are more oil reserves to be discovered (one billion barrels were found in 1994 alone) and staggering profits to be made in the process. Royal Dutch Shell, through SPDC and its new baby, SNEPCO, aims to be in Nigeria for a very long time.

The Nigerian Oil and Gas Industry

Any Nigerian schoolchild will readily tell you that Oloibiri, a small village in the Niger Delta, is where the first oil well was struck by Shell in 1956. It is one of the first lessons they learn in primary school. But the story of oil exploration in Nigeria actually began in 1908 when a German company,

Nigerian Bitumen Corporation, set up shop. It was, however, a joint venture between Shell and British Petroleum (BP) that discovered the Oloibiri reserves in 1956. Commercial exploitation began two years later.[31] Four decades after Oloibiri, Nigeria is ranked as the thirteenth-largest producer of oil in the world, accounting for almost 3 percent of entire global production. Nigeria is also a leading producer of natural gas, with reserves of 13.3 trillion cubic feet, 2.4 percent of total world reserves. There are seventeen oil companies producing from about 150 oil fields presently. Ninety percent of these fields are located in the Niger Delta.[32]

Oil is the pivot on which the Nigerian economy revolves. Oil sales currently provide over 95 percent of the country's export earnings and also account for some 25 percent of Gross Domestic Product. So dependent has the country become on the oil fields of the Niger Delta that it is often said that the entire country would grind to a halt, at least temporarily, were the oil to suddenly vanish. It is not easy to estimate exactly how much Nigeria has earned from oil exports to date. This is because of the endemic corruption that has plagued both the industry and the government. According to an assessment carried out by the International Monetary Fund, the country earned a total of $65.6 billion from the oil fields of the Niger Delta between 1985 and 1992.[33] The government-owned Nigerian National Petroleum Company has said the country earned $101 billion in oil revenue between 1958 and 1983.[34] The government's figures are not reliable, however, as the Justice Irikefe Panel that examined the books of the NNPC in 1980 discovered—accounts were not being properly kept.[35]

The price of Nigerian crude has always fluctuated in response to the vagaries of the international market. From an all-time high of $37.20 in 1980 for a barrel of Bonny Light, the country's premier crude plunged to $14.60 in 1986. The Gulf War, which broke out in 1990, stimulated oil prices again and Bonny Light sold for $24.30 a barrel. But the recovery was short-lived, as the bottom fell out of the international oil market in 1994, forcing prices down to $16.20. The market picked up in the third quarter of 1996, with Bonny Light commanding $26 a barrel.[36] Oil prices dipped to $10 to $12 per barrel a year later before swinging up yet again in 1999.

While prices have been erratic, what is beyond doubt is that the oil fields of the Niger Delta brought great wealth to the country, especially during the boom years of 1972–79, when production peaked at 2.3 million barrels a day. The oil boom also brought with it massive corruption, misappropriation of funds, and a craze for imported luxuries, especially among the military rulers, senior civil servants, and the urban elite. The gap between urban and rural Nigeria widened even further—at one end luxury motorcars and palatial mansions and all the other "good things" of life, and at the other, poverty, misery, and disease. A World Bank policy paper on Nigeria estimates that over $68 billion was siphoned from the country by successive military dictators and their civilian collaborators between 1972 and 1989.[37] The people of the Niger Delta saw very little of the oil proceeds.

The Big Players

Nigerian crude is very popular among oil companies and buyers alike because it is very light and has low sulfur content. It is therefore highly sought after by refineries in Europe and the United States, where there are very strict rules guiding environmental pollution. Nigeria's Bonny Light and Forcados burn easily in the process of refining and discharge minimum waste into the atmosphere.

There are six major oil companies currently operating in the country. Shell, along with NNPC, Elf, and Agip, produces about 42.2 percent of Nigeria's total output; Mobil 21.2 percent; Chevron 18.6 percent; Agip 7.5 percent; Elf in partnership with NNPC 6.1 percent; and Texaco 2.6 percent. There are other oil companies that produce oil in the country, including indigenous ones. These are: Ashland (U.S.), Deminex (Germany), Pan Ocean (Switzerland), British Gas (Britain), Sun Oil (U.S.), Conoco (U.S.), BP (Britain), Statoil (Norway), Conoil (Nigeria), and Dubril Oil (Nigeria). Between them, these ten oil operators account for 1.7 percent. The United States is Nigeria's main buyer, taking about 40 percent of annual sales. Spain comes second with 14 percent. Other major buyers are South Korea, India, France, Japan, China, Taiwan, the Philippines, and Thailand.[38]

The Nigerian National Petroleum Company (NNPC)

Nigeria joined the oil cartel, the Organization of Petroleum Exporting Countries (OPEC), in July 1971 primarily to safeguard her interests in the international oil market and ensure, along with the other members, that the major consumer nations of Europe and North America do not force down oil prices by playing one producer against the other.[39] OPEC regulates annual oil production, and by doing so influences international oil prices through such mechanisms as production quotas and ceilings that all members are obliged to obey. The federal government had established the Nigerian National Oil Corporation (NNOC) by decree the previous May in response to OPEC Resolution No. XVI.90 of 1968, which urged member countries to "acquire 51 percent of foreign equity interests and to participate more actively in all aspects of oil production."[40] Eight years later, in August 1979, the military administration of General Olusegun Obasanjo nationalized BP's share equity in the Shell-BP joint venture in an attempt to force the British government, the majority shareholder in the company, to take a firmer stand on the issue of sanctions against the racist government of Ian Smith in Rhodesia (now Zimbabwe), and also announced that the NNOC, renamed the Nigerian National Petroleum Company (NNPC) in 1977, would manage BP's assets in the country.

British Petroleum had participated with Shell on a fifty-fifty basis to develop the first oil fields in the Niger Delta in the 1950s. By 1961, along with Shell, it had acquired four Oil Exploration Licenses (OELs) from the Nigerian government. Before BP's share of the Shell-BP partnership was fully nationalized by the Nigerian government in 1979, the company's assets in the country included eighty oil fields, including pipelines and storage facilities, a 20 percent share in a refinery in Port Harcourt, a 60 percent stake in British Petroleum Nigeria Marketing Company, and a further 20 percent in the Shell-BP-NNPC oil-producing venture.

Like Shell, BP was also implicated in the controversial dealings with apartheid South Africa, and was accused by the Nigerian government of deliberately breaking the oil embargo against the apartheid state by exchanging its North Sea crude for Indonesian crude from Conoco, resulting in some of the oil being shipped to South Africa. This, along with the Nigerian government's plan to pressure Britain on the Rhodesia issue, led to the nationalization of BP's assets in the country. BP returned to Nigeria in 1991 through a new partnership with Statoil, the Norwegian government-owned oil company. An exploration license was offered the BP-Statoil combine the same year, and two years later it signed a production-sharing venture agreement with the NNPC.

NNPC enjoys a privileged position in the Nigerian oil industry. It is not involved in oil exploration or production, but has the lion's share (55 to 60 percent) in all joint ventures in the country. NNPC has a 55 percent stake in Shell Petroleum Development Company Ltd., 58 percent in Mobil Producing Nigeria Ltd., 58 percent in Chevron Nigeria Ltd., 60 percent in Elf Petroleum Nigeria Ltd., and another 60 percent in the Texaco Overseas (Nigeria) Petroleum Company.[41] Recently, though, the NNPC has been taking tentative steps toward developing its own oil wells independent of the foreign oil companies with which it is in partnership. The NNPC owns and manages the country's four refineries, two of which are in Port Harcourt, the others in Warri and Kaduna in the north. It has a joint venture partnership with several petroleum products marketers in the country, among them African Petroleum, National Oil, and Unipetrol, and it regulates the distribution and sale of petrol and allied products by issuing licenses, franchises, and permits to dealers all over the country.

Killing the Goose

The Nigerian oil industry has, over the years, been characterized by massive corruption and shoddy management. Refined petroleum products are cheaper in Nigeria than in neighboring West African countries, and it had been the standard practice for the army generals and their cronies, when they were not looting the state treasury directly, to smuggle gasoline out for sale in the subregion at great profit. About 30 percent of all refined petroleum products leave the country through this illegal route, and this causes severe shortages in parts of the country. Industry sources said some 100,000 barrels of oil were smuggled daily overland into Cameroon, Benin, and Niger in 1993.[42] Dealers also collude with senior government officials to hoard the product and force up prices.

Corruption and misplacement of priorities have also been the bane of the country's four refineries. Not only are they poorly maintained, high officials sometimes deliberately sabotage operations and even burn down facilities in an attempt to cover up their misdeeds and create the opportunity for repair work, which in turn is contracted out to incompetent cronies at grossly inflated rates. Production is erratic, even in the second refinery in Port Harcourt, which was built only in 1989. Between them, the four refineries manage to meet only half of their production capacity of 455,500 barrels a day, not anywhere near meeting local consumption needs. The federal government regularly imports petroleum products to bridge the shortfall, but even this has been turned into a racket. The license to import oil products is the exclusive preserve of senior army generals (serving and retired) and ranking civil servants, using a number of front companies. They buy cheap in European markets, sell dear to Nigerian consumers, and pocket the excess profit.

The Nigerian people, exploited and subjected to the agony of lining up in gas stations for weeks on end, have always resisted attempts by the soldiers to increase the pump price of gas and other petroleum products. The government, under pressure by the IMF, claims that Nigerian gas is the cheapest in the world and that it is heavily subsidized. It wants to remove the "subsidy" so that, according to it, Nigerians will pay the proper economic price for the product, and government will in turn earn the revenue needed to properly maintain the refineries for the benefit of all. Labor leaders and other critics of the government's structural adjustment policies have, however, contended that there is in fact no subsidy on petroleum products, that Nigerian workers are grossly underpaid, and that there will be more than enough money to maintain the refineries if only the people in power are willing to stop looting the billions the country earns every year from the export of crude.

In the face of considerable civil opposition, the government has been increasing the price of petroleum products in installments, the most recent of which was effected in June 2000.[43] This has brought untold hardship to Nigerian workers and peasants, forcing up the price of food and transport. This development, in addition to the public resentment and anger following the brazen annulment of the June 12, 1993, presidential election results by the Babangida junta, finally pushed the country's oil workers to the wall. Their two umbrella organizations, NUPENG and PENGASSAN, declared a strike in June 1994 that for ten weeks brought the country to a standstill. Several of their leaders, including Ovie Kokori, General Secretary of NUPENG, were detained in August 1994 when the striking oil workers were eventually forced back to work and their union secretariats taken over by government lackeys.

In June 2000, one year after he assumed office as president following the end of military rule, Olusegun Obasanjo unilaterally increased the pump price of gasoline by fifty percent, claiming his government needed the extra revenue to deliver key social services to the people. The move was stoutly resisted by the umbrella workers organization, the Nigerian Labor Congress, whose leader, Adams Oshiomole, called out workers on a strike that crippled commercial life in the country for three days. President Obasanjo was forced to back down, and a compromise arrangement was worked out between the two parties after tense negotiations, increasing the price of gasoline only marginally while leaving the price of kerosene, the major cooking fuel in the country, untouched.

Shell*ing* Nigeria

Shell has done very well for itself in a Nigeria repressed, brutalized, and looted in turns by successive military dictators. To outsiders, the company projects the image of neutrality in the quicksand of Nigerian politics, a wise and benevolent patriarch towering above the chaos and corruption that government and public life has become in Nigeria since the civil war. The reality, however, is that the multinational has quietly and unobtrusively worked its way to the epicenter of power over the years. It enjoys cordial relations with the soldiers and politicians in power, in a symbiotic relationship sustained by a mutual desire to control the Niger Delta and exploit the oil. As the American environmental pressure group Project Underground pointed out in 1997 (when General Abacha was still in power), "Shell supplies fully half of the income to a brutal regime bent on suppressing dissent."[44]

It is a measure of how powerful Shell has become in Nigeria and the extent to which its business interests had merged with the designs of one of the most brutal and corrupt regimes in the world that Dr. Owens Wiwa, brother of the murdered environmentalist, told journalists he had a meeting with the former managing director, Brian Anderson, in his home in Lagos in May 1995, and that Anderson said he could effect Ken Saro-Wiwa's release from detention but would only do so if MOSOP called off its international campaign against his company.[45] Ken Saro-Wiwa rejected the offer out of hand, and a few months later he was hanged. Shell has admitted that Anderson did meet with Dr. Wiwa but disputes the latter's account of what transpired between them.

The multinational maintains its own private police force, imports its own arms and ammunition, and at least in two instances has admitted payments to the Nigerian military.[46] Officials of Project Underground have also drawn attention to the existence of three separate Shell armories in Bonny, Warri, and Port Harcourt—all in the Niger Delta. There are pump-action shotguns, automatic rifles, and revolvers in these armories.[47] Shell maintains that the weapons stored there are for the police officers assigned to the company by the Nigerian government and that it has an inventory of all the ammunition and would know immediately if any is missing.[48] Project Underground, however, spoke to former Shell police staff members, who explained that there was "no account of bullets," and that while these bullets were recorded, Shell invariably had more bullets in its armories than was officially in its books.[49]

These Nigerian police officers assigned to Shell are referred to as Shell Police by local people in the Niger Delta. They are paid directly by the company instead of the Nigerian government, and take their instructions from Shell officials. Shell Police is something of an elite force, not unlike General Babangida's now disbanded National Guard. The officers, unlike their counterparts in the regular force, receive free accommodation, transport, meals, medical services, and regular lump-sum payments, at least double the government rate, courtesy of Shell. Sometimes, especially when they are engaged in undercover operations on behalf of the oil company, they move about in plainclothes.[50]

Shell Police has four units: Operations (OPS), whose primary duty is to provide security at company installations; Administration, which provides administrative support for the operations of the force; Intelligence and Investigations, whose members investigate community compensation claims in case of oil spills and usually operate clandestinely; and the Dogs and Arms Section, which supervises the armories and the specially trained dogs Shell Police officers use in their work.[51]

While Shell officials insist that "the policemen concerned are specifically assigned to SPDC for the sole purpose of carrying out such guard duties," the local communities have accused them of brutally suppressing peaceful protests and using financial inducements to divide the community whenever there is an oil spill, so they cannot present a common front and successfully press for compensation.[52] Four former members of Shell Police who spoke with Project Underground in April 1997 testified that Shell officials would give them "service money," which they used to gather intelligence and bribe and befriend villagers wherever there was an oil spill. These villagers would then instigate conflict in the village over competing claims for money, a situation Shell would subsequently exploit, claiming that it would not pay any compensation since the community was divided on the issue of who would get what. These officers also talked about a special "strike force," which they claimed was deployed to suppress community protests, armed with automatic weapons and tear gas canisters.[53] Dr. Owens Wiwa and other MOSOP activists stated that members of Shell Police, accompanied by military troops, were ferried by the oil company's helicopters and boats to attack Ogoni villages.[54] Other communities in the Delta have also recounted similar experiences.

As early as 1991, a former managing director of Shell Nigeria, Phil Watts, retained five members of the notorious Nigerian mobile police force for his personal protection.[55] Members of Shell Police have also been implicated in the abduction of Ogoni people.[56] The late Claude Ake described this disturbing trend as the privatization of the Nigerian state by Shell officials. Said Professor Ake, "The privatization of the state is evident in the swarm of policemen and -women in Shell residential quarters and offices supposedly securing Shell, the presence of armed troops in the operational bases of the company, and in the prerogative of Shell and other oil companies to call on the police and the military for their security."[57]

Shell has had on its board of directors some very influential Nigerians, among them Ernest Shonekan, who was head of the illegal Interim National Government before it was sacked by General Abacha in November 1993. Shonekan has since left the board. In November 1996, Abacha appointed him chairman of the 173-member Vision 2010, a committee set up by the junta ostensibly to chart a new economic direction for the country but which was dismissed by perceptive critics as yet another of the elaborate ploys that the late dictator had put in place to perpetuate himself in power. Conveniently, Vision 2010 happens to be based on a scenario model developed by Royal Dutch Shell.[58]

FOUR

A Dying Land

Then there was a big spillage. The oil just came out of the ground and it was more than they could cope with. It circulated in the rivers and many fish died, and where it touched the land, food crops died and the land became infertile.

Princess Irene Amangala of Oloibiri village[1]

Ecology and Resources of the Niger Delta

The Niger Delta is the most extensive lowland forest and aquatic ecosystem in West Africa, and has very high concentrations of biodiversity.[2] It is a complex tangle of creeks, streams, and swamps formed by the Niger River as it divides into six main tidal channels just before it enters the Atlantic Ocean. The Delta floodplain consists of accumulated sediments deposited by the Niger and Benue rivers and has four major ecological zones:

1. Coastal sand barrier islands, mainly along the Delta coastline: This ecozone has four subecozones and is characterised by ridgetop tropical forest, trough freshwater swamp forest, brackish water swamp forest, and sandy beaches.[3]
2. West African lowland equatorial monsoon: This ecozone is marked mainly by high and low water table, vast stretches of floodplain, and riverine swamp.
3. West African freshwater alluvial equatorial monsoon: This is a levee forest area and is also marked by palm swamp and seasonal swamp. There are white-water and black-water floodplain, lakes, and rivers. The fresh-

water swamp forests of the Delta cover an area of 4,500 square miles, clearly the most extensive in west and central Africa. The forests are seasonally flooded, and while it is difficult for farmers to cultivate in this area, it has nevertheless been subjected to some logging, leading to gradual degradation.

4. West African brackish-water alluvial equatorial monsoon: This ecozone is dominated by mangroves. It is also the area of transition between mangroves and freshwater alluvial equatorial monsoon. Nigeria's mangroves are the largest in Africa, and over half of this is in the Niger Delta. There is an extensive network of creeks, and the mangroves grow in such profusion and compactness as to make human penetration difficult. This ecozone is therefore one of the least degraded in the Niger Delta.

The Niger Delta is low, flat terrain, straddling five degrees north of the Equator. It is fairly extensive, stretching into the Gulf of Guinea and forming the Bight of Biafra in the east and the Bight of Benin in the west. The climate here is tropical hot monsoon. Rainfall is very high, averaging an annual 117 to 175 inches. Discharges from the rivers and creeks peak during the rainy season (July–September), and because the soils are poorly drained, over 80 percent of the Niger Delta is seasonally flooded. Erosion is also a recurring phenomenon. When the flood eventually recedes during the dry season between December and January, the water channels that fan out over the Delta leave swamps and small lakes in their wake. Average monthly temperatures hover around seventy degrees.[4]

Before dams were built in the Niger, beginning with the Kainji Dam in 1968, the Delta was a dynamic and self-regulating ecosystem. There was a fine balance between the constant flooding, erosion, and sediment deposition. However, with the construction of dams upstream, it has been estimated that about 70 percent of the sediment transport from the Niger and its tributaries has been lost, and this, combined with oil production, has severely disrupted the natural equilibrium of the Niger Delta ecosystem. Sediment deposition is gradually rising again, though, due mainly to the accumulation of silt in the dams, which has led to a decrease in the capacity of the reservoirs to obstruct river flow.[5]

In terms of natural resources, the Niger Delta is one of the world's richest areas. Apart from its substantial oil and gas deposits, there are extensive forests, abundant wildlife, and fertile agricultural land where rice, sugarcane, plantain, beans, palm oil, yams, cassava, and timber are cultivated. The Delta is also famous for its fish resources. It has more freshwater fish species than any other coastal system in West Africa.[6] Indeed, three-quarters of the fish caught in the subregion are bred in the mangroves of the Delta, which have been described as the third largest and the most discrete in the world.[7] Mangrove trees, which grow tall and healthy on the creeks and near riverbanks, provide protective barriers for the country's coast and are also a source of medicine, fruit, and raw material for such cottage industries as weaving, wood carving, and rope making. The biodiversity of the Delta is enormous. The World Bank has drawn attention to its importance as home to a great variety of threatened coastal and estuarine fauna and flora, and to the need for preservation of the biodiversity of the area because of its rich biological resources.[8]

Shell in the Niger Delta

Shell has been described as a major polluter of the environment on the one hand, and a busy propagator and purveyor of technical fixes for its transgressions on the other.[9] It is therefore not always easy to penetrate the elaborate "environmentally friendly" facade erected by the company's green lobbyists and spin doctors to the ogre that is polluting and despoiling the world's fragile ecosystems.

Shell's oil exploration and production activities in the Niger Delta are almost entirely based on land.[10] This means that the bulk of the company's operations—ninety-four producing oil fields scattered through well over 12,000 square miles, eighty-six production stations, and more than 3,700 miles of pipeline—take place in the same ecosystem inhabited by the various communities of the area, including the flora and fauna.

In the course of exploring for oil in the Niger Delta these past four decades, Shell, contrary to what the various public relations agencies and consultants in its employ have been striving to sell to the international community, has not only radically disrupted the ecological balance of the area, but through negligence and cynical indifference has orchestrated a vicious ecological war—a war whose victims are a hapless people and the land on which they have lived and thrived for centuries. The report submitted to the World Conference of Indigenous Peoples on Environment and Development during the Rio Earth Summit in June 1992 by the kings, chiefs, and community leaders of the Niger Delta told the harrowing story of the still-ongoing ecological war waged by Shell in the Delta:

> Apart from air pollution from the oil industry's emissions and flares day and night, producing poisonous gases that are silently and systematically wiping out vulnerable airborne biota and otherwise endangering the life of plants, game, and man himself, we have widespread water pollution and soil and land pollution that respectively result in the death of most acquatic eggs and juvenile stages of life of finfish and shellfish and sensible animals (like oysters) on the one hand, whilst, on the other hand, agricultural lands contaminated with oil spills become dangerous for farming, even where they continue to produce any significant yields.[11]

In 1983, long before Shell's activities in the Niger Delta made international headlines, officials of the Inspectorate Division of the Nigerian National Petroleum Company had sounded an alarm over what the oil exploration and production activities of Shell and the other foreign oil companies were doing to the Delta environment. Said the NNPC report, "We witnessed the slow poisoning of the waters of this country and the destruction of vegetation and agricultural land by oil spills which occur during petroleum operations. But since the inception of the oil industry in Nigeria more than twenty-five years ago, there has been no concerned and effective effort on the part of the government, let alone the oil operators, to control the environmental problems associated with the industry."[12] Nigerian government officials are noted for their cautious use of words. They must have been sufficiently outraged at the spectacle of environmental destruction that assaulted their eyes in the Niger Delta to state the condition of things in such stark terms. Seventeen years after they made this report little has changed.

If anything, the pace of environmental pollution has accelerated as Shell intensifies its oil production and exploration activities in the area, pushing up its production target to one million barrels of crude a day. In the process of extracting the oil, adequate consideration is not given to the over seven million people who live in the area, and the impact of the company's operations on their environment and their way of life. Indeed, since Shell set up its first oil rig in Oloibiri in 1958, not a single satisfactory Environmental Impact Assessment (EIA) has been conducted and made public in the Niger Delta before operations commence, to determine what potential harmful effects such activities are likely to have on the area and how to avoid or at best minimize them.[13] Shell vigorously denies this, claiming that it has been conducting EIAs for its operations in the Niger Delta since 1982—but the company has not been able to back its denial with adequate evidence. The two EIAs that the company claimed were commissioned for a major pipeline project in the Delta turned out to have been conducted well after the project had commenced.[14] Body Shop International commissioned Environmental Resources Management, an environmental consultancy firm, to examine the two EIAs, and it discovered that they were vague in conception and inadequate in implementation. Said the Environmental Resources Management report, "Both documents [the Shell-commissioned EIAs] refer to Shell's Oil Spill Contingency Plan as a major mitigative measure, but there is no clear indication that an effective contingency plan, customized to account for specific local environmental sensitivities, in fact exists." The report also added that "there is little evidence that SPDC have been involved in the EIA process, that they acknowledge the potential impacts of their pipeline operations, and that they have taken ownership of the mitigation measures necessary to minimize potential impacts."[15]

All available evidence suggests that Shell's destruction of the Niger Delta is informed by near-total disregard for the welfare of the local people.[16] Why else would the same company go to great lengths to conduct rigorous and extensive EIAs for its operations in Europe and North America and refuse to do the same in the Niger Delta? Consider for a moment this report on seventeen different EIAs that Shell conducted for a pipeline project in Scotland before a single hole was dug: "A painstakingly detailed Environmental Impact Assessment covered every meter of the route, and each hedge, wall, and fence was catalogued and ultimately replaced or rebuilt exactly as it had been before Shell arrived. Elaborate measures were taken to avoid lasting disfiguration, and the route was diverted in several places to accommodate environmental concerns."[17] Clearly, what is good for the people of Scotland is not considered necessary or desirable for the communities of the Niger Delta, from whose land Shell has extracted billions of dollars' worth of oil since 1958.

The net effect of Shell's environment-destroying operations in the Niger Delta is an ecosystem so mangled, raped, and denuded that the area has been labeled the most endangered delta in the world.[18] The carnage is total and all-pervasive—high-pressure pipelines that crisscross farmlands and even house backyards, well blowouts, and discharge of waste and flares that light up the skies twenty-four hours a day and poison the atmosphere with lethal gasses. David Moffat, an environmental consultant with the World Bank, has estimated that since it began operations in the Niger Delta, Shell has destroyed a substantial portion of the mangrove forests in Rivers and Delta states alone, in the process also exposing this otherwise discrete ecosystem to further degradation by hunters and loggers.[19] The company said it had 890 production wells in operation in the Niger Delta as of May 1997.[20] Blowouts occur regularly in some of these wells, particularly in the flow stations, polluting farmland and rivers and creeks with oil, and destroying flora and fauna in the process.

The World Bank estimates that oil companies in Rivers and Delta states spill about 9,000 cubic feet of oil in three hundred major accidents yearly.[21] On its part, Shell says it spilled an average of 7,350 barrels of oil a year between 1989 and 1994, and that a total of 221 spills occurred in the course of its operations during the period. However, as Greenpeace has pointed out, these figures do not include the large number of supposedly "minor" spills that occur daily but which Shell usually did not take into account in its rough estimation.[22] Besides, Nigerian crude is very light and is quick to evaporate, making it difficult to assess the precise volume and spread of spills when they occur. The World Bank therefore argues that the actual figure of oil spills in the Niger Delta every year is actually about ten times the official estimate.[23]

These oil spills occur because the bulk of Shell's pipelines through which the oil leaks are rusty, obsolete, and poorly maintained. Some Shell pipelines and sundry installations in the Niger Delta have not been replaced since they were put in place in the 1960s.[24] The result has been an increase in the rate and volume of oil spills as Shell accelerates production activities, subjecting old and weary pipelines to pressure they are no longer able to handle. They crack and buckle, spewing oil into the surroundings. The testimony of J. P. Van Dessel, Shell Nigeria's former head of environmental studies, best sums up this ecological nightmare: "Wherever I went, I could see . . . that Shell's installations were not working cleanly. They didn't satisfy their own standards, and they didn't satisfy international standards. Every Shell terrain I saw was polluted, every terminal I saw was contaminated." Van Dessel was so outraged at Shell officials' indifference to this shocking scenario that he threw in his resignation letter in December 1994, two years after he took up his post.[25]

Nigeria also has the dubious honor of being the world leader (including all OPEC countries) in flaring gas brought up with oil in the drilling and extraction process. The World Bank estimates that 87 percent of all associated gas is flared into the Niger Delta atmosphere by oil companies operating in Nigeria, compared to 21 percent in Libya and 0.6 percent in the United States.[26] Shell officials said the company flared an average 40 billion square feet of gas every year between 1991 and 1994, and according to these figures, the World Bank has estimated that 80 billion cubic feet of gas is flared in the Niger Delta yearly.[27] Shell's operation in Nigeria, according to Geoffrey Lean, the leading British environmental journalist, makes the company one of the biggest contributors to global warming. The company's gas-flaring installations are like its pipelines—old, poorly constructed, and in some cases ill-maintained—and as a result they emit "far more pollution than Britain's twenty million homes put together."[28] Interestingly, the percentage of gas flared in the Netherlands, where Royal Dutch Shell has its international headquarters, is zero.[29]

For eight years the World Wide Fund for Nature (WWF), following entreaties from concerned Nigerian scientists and conservationists, secretly lobbied Shell to clean up its operations in Nigeria and ensure that the amount of gas flared in the course of its operations is substantially reduced. Shell, however, consistently rebuffed these pleas. Faced with this obduracy, the WWF went public in December 1995, denouncing Shell's operations in the Niger Delta. After a tour of Shell's installations in Nigeria, an appalled Clive Wicks, the head of WWF-U.K.'s international program, declared: "Traveling in the area is like flying over Dante's inferno. Wherever you look you see these goddamned flares."[30]

Bull in a China Shop

We can get a fair picture of the process by which Shell ravages the most endangered ecosystem in the world by considering a typical day in the life of a Shell worker as he sets out to explore, drill, and transport oil in the swamps, creeks, and mangrove forests of the Niger Delta. The activities of this typical Shell worker as he sets out each morning have been likened to those of a bull running amok in a china shop, trampling, slashing, and kicking everything in its way to smithereens.[31] The process by which crude oil is found and put to commercial use goes through several stages, each of them a lethal blow to the tender underside of the Delta ecosystem.

Seismic Survey

To find oil in the Niger Delta, Shell engineers conduct geological analyses of the area through extensive seismic surveys.[32] Since the bulk of the company's operations are on land, seismic crews operate in the mangrove swamps, freshwater creeks, and, in some cases, agricultural land. Line cutting is the first stage in seismic surveys. An area of land is identified and demarcated. Then the path through which the seismic waves will travel (the shot and receiver lines) is cleared of flora and other likely impediments, to a minimum width of about three feet. This is done with machetes and involves cutting down huge swaths of trees and other vegetation. When the clearing is completed, deep holes are dug, which are flushed with standing or creek water, and explosives (usually dynamite) are placed in them, alongside detonation equipment. The explosives are then detonated and the signals recorded at a central recording station using the appropriate instruments.

Since 1986, Shell workers have employed the three-dimension survey technique (3D Survey), which is more cost effective than the old two-dimension survey. The 3D survey crews are very large, sometimes numbering 1,200 workers. According to Van Dessel, Shell has cut 37,000 miles of 2D seismic lines since it began operations in the Niger Delta, more than 24,000 miles of these in the mangroves. The 3D lines total 19,460 miles, with 10,790 in the mangroves. Van Dessel also estimates, based on the current one-meter-line width allowed for seismic surveys in Shell's Eastern Division operations, that approximately 35 square miles of land was cleared for the company's seismic surveys; 22 square miles of this is mangrove.[33]

Shell operated on its own in the Niger Delta from 1956 to 1977, when it teamed up with NNPC, Elf, and Agip to establish SPDC. And for all those twenty-one years, not a single adequate Environmental Impact Assessment was conducted to determine the potential havoc seismic surveys could wreak on the Niger Delta ecosystem. During the oil company's seismic activities, forests are invaded and cleared, and animal species endemic to that particular habitat are either expelled or killed. Bush clearing during the line-cutting stage also makes the forests accessible to humankind, a process that further accelerates the destruction of rare animal species. It is in the mangrove swamps of the Niger Delta that the ravages of Shell's seismic activities are most noticeable. Here the aerial roots of tall mangrove trees are mauled and ravaged, and it takes them over three decades to regenerate—that is, if the area is not disturbed by renewed oil production activities.

The detonation of explosives also disrupts the natural structure of the soil, and where the holes are not properly dug, they disintegrate and cause a crater, further disfiguring the landscape. Alien chemicals are also introduced into the soil after the explosives are detonated. As Moffat and Linden have pointed out, the Niger Delta ecosystem has one of the highest concentrations of biodiversity in the world.[34] And given that an approximate twenty-two square miles of mangrove has been cut by Shell in its Eastern Division alone in the course of its seismic operations, a considerable amount of fauna and flora have been destroyed, expelled, or damaged beyond repair during the period.

Drilling

Shell's drilling operations are in four stages: preparing the drilling site, exploration drilling, production testing, and transport. Before the drilling for oil commences, access roads are cut through to the drilling site. The site proper is also cleared of all vegetation. Then the site and access channels are dredged to a stipulated depth and width using bucket dredgers and a dredging barge. When this is completed, the drilling rig is towed to the site by tugboats. Where there is no Shell flow station nearby, a flare pit is provided for production tests to determine the commercial feasibility of the well. Shell's engineers use Water Based Muds (WBM) to drill the wells rather than Oil Based Muds. Production testing usually lasts a few days but could extend to several weeks in some cases. Oil and gas brought to the surface during this period is either sent to the nearest flow station through pipelines or is flared right there on the site. Shell's drilling operations are usually massive activities, involving men, drilling equipment, and vehicular transport like boats, four-wheel drives, and helicopters.[35]

As in seismic surveys, trees and other vegetation cut down in the process of site preparation for drilling result in serious damage to the Niger Delta ecosystem. Dredging is particularly harmful to the Delta ecology. Apart from land that is lost in the process of the dredging proper, dredged material is dumped on either side of the canals, and because this waste is usually high in organic content and turns acidic in the process of oxidation, it destroys the ecology of the surrounding area where it is dumped. After a while some of the dredged material is washed back into the creeks or canal constructed in the process of site preparation, and this tends to increase sedimentation in the creeks and the turbidity of the water, leading to a significant reduction in the penetration rate of sunlight.[36] Phytoplanktons that depend on sunlight to thrive and reproduce are thus harmed. Water turbidity also makes it difficult for such birds as kingfishers to see fishes in the water clearly, and has also been known to severely affect certain species of fish that do not thrive in muddy water. Sedimentation in creeks leads to the destruction of benthic fauna and other sensitive aquatic creatures. The Water Based Muds discharged into the creeks have a high starch content. All available oxygen is used up due to increased bacterial activity, with the result that aquatic flora and fauna that depend on this important energy source are either expelled or wiped out completely. Shell workers dispose of their drilling waste in waste pits. Indeed, the entire Niger Delta is littered with these waste pits, and they sometimes overflow during the rainy season, discharging mud high in salt content into surrounding farmlands.

Shell's production-testing operations is usually a very noisy affair. The concentration of large bodies of men in a hitherto serene habitat, the noise they make as they tow the rig through the canal and operate other drilling equipment, and the noise generated by the rig itself as drilling commences, not only scare away wildlife in the vicinity but also disturb the peace of local people. Sometimes unburned gas and oil mixed with production water overflow from Shell's flare pits in the course of test drilling, polluting water and soil in the vicinity through discharge of hydrocarbons. Blowouts also occur during production testing, further damaging the environment.

Oil and Gas Production

The River Nun in the eastern Delta was once famous for its serene beauty and was celebrated by the Ijo poet Gabriel Imomotimi Okara in his classic work, "The Call of the River Nun." Today, however, this river, a tributary of the Niger, has been reduced to an ugly caricature of its former self. Gone are the beautiful beaches and the somnolent waves at high tide that once seduced the famous poet. Thanks to Shell's activities in the area, and the construction of extensive oil and gas production infrastructure, the hydrological character of the once beautiful River Nun, particularly the area around the company's oil field at the Nun River flow station, have been radically altered for the worse, perhaps forever.[37]

Elsewhere, waste generated through production activities is discarded in an environmentally hazardous manner. Van Dessel and WWF officials who toured Shell's production facilities have graphically chronicled this aspect of the company's operations, describing it as primitive, inadequate, and highly lethal. In most cases, Shell's waste-treatment program is as obsolete and inefficient as are its facilities. Waste is burned either in flare pits or in a "very primitive barbecuelike incinerator"—to use Van Dessel's words.[38] The other options are burial in waste pits for oily degradable and undegradable waste, and injection into oil pipelines where small quantities of hydrocarbon-based toxic chemicals are involved.

It has been pointed out that none of these three methods of waste treatment is satisfactory from an environmental and health perspective.[39] Buried oily waste tends to contain harmful chemical components and pollute groundwater when it seeps into it. Nor are Shell's incineration facilities famous for their efficiency. A lethal cocktail of unburned hydrocarbon, soot, and heavy metals is emitted into the Niger Delta atmosphere, contributing to global warming and causing the local inhabitants considerable discomfort.[40] A proposal to purchase a new industrial incinerator that would burn the company's waste through an efficient and controlled process was postponed by senior company officials.[41] Another cause for worry is the manner in which the company's officials dispose of crude leaked in flow stations, and also oil recovered during spill cleanup operations. This oil is burned in the flare pits in the flow stations, sending poisonous gases into the atmosphere. Said Van Dessel, "What is happening in Nigeria is if there is an oil spill then the spill will be cleaned up, in the sense that it is more or less removed using buckets and spades . . . Well, that's the level of the operation. So what's usually being done is they dig a hole in the middle of the oil spill, and the oil is collected with buckets and spades and collected in this pit, and that's it."[42] Abandoned oil wells and waste pits dot the Niger Delta landscape, and nobody is talking about cleaning them up. Yet.[43]

Oil, Oil Everywhere

The bulk of Shell's production facilities, especially its pipelines, are rusty and obsolete. It was, and continues to be, Shell Nigeria's stated policy to replace flow lines in swampy areas every ten to fifteen years. However, in 1995 Van Dessel stated, "At the moment there is still a backlog of older pipelines with high leakage frequencies."[44] As of this writing, pipes continue to burst, ruining lives, fishing creeks, and farmlands in the Niger Delta. Consequently, spillage is a regular occurrence in the Niger Delta—and this occurs in the immediate vicinity of oil wells during massive blowouts, spillage from leaking pipelines, and in the terminals where oil is separated from the production water. The number of oil spills registered in the course of Shell's operations has been increasing in recent years, as the company steps up oil production in the Niger Delta.

While company officials claim that a good part of this spillage is caused by the Ogoni and the other oil-producing communities that sabotage Shell facilities in order to claim compensation, they have not been able to supply credible evidence—as a July 1996 British Advertising Standards Authority ruling against Shell has demonstrated.[45] Claude Ake, who resigned from the Shell-sponsored Niger Delta Environmental Survey in the wake of Ken Saro-Wiwa's murder in November 1995, also dismissed Shell's claim of sabotage as irresponsible, arguing that it is in fact the local communities who go out of their way to clean up the oil spills. Said Professor Ake, who died in a mysterious plane crash in Nigeria in October 1996, "Nobody can say that most of the pollution in Ogoniland is caused by sabotage. In fact, as far as I know, what the Ogoni have tried to do is to put out the flares, which is something that importantly reduces pollution. I think that is the kind of propaganda that the oil companies are putting out in order to discredit those who are trying to do something about the environment."[46]

Several postimpact studies conducted in areas that suffered spillage due to Shell's production operations in the Niger Delta, particularly the Ebubu spill, clearly indicate that they adversely impact on the lives of the local communities and the flora and fauna.[47] Drinking water is severely polluted, mangrove trees and food crops are smothered to death, and fish, oysters, and other water creatures are destroyed in contaminated streams, swamps, and creeks.

More gas is flared in the course of Shell's operations in Nigeria than in any of the other hundred countries in the world where the multinational is involved in oil exploration and production activities. This is so because Western oil companies operating in Nigeria find it economically expedient to flare nonassociated gas right there on the spot in the flow stations rather than incur the expense of putting in place facilities to reinject the gas back into the wells or collect it for commercial use. Shell Nigeria flared an average of 32 billion cubic feet of gas into the Niger Delta atmosphere every year between 1991 and 1994.[48] The World Bank also estimates that Shell's gas flares in the area released 39 million tons of CO_2 and 13 million tons of methane into the atmosphere in 1994 alone.[49] The marginal damage caused by these carbon emissions has been put at $7.50/ton.[50] The flare stacks in Shell's flow stations are poorly designed, spewing unburned gas into the atmosphere and sometimes directly at homesteads.[51] Local communities have borne the brunt of these gas flares. The roofs of their houses are severely corroded, the heat generated by the gas flares leads to reduced crop yield, and the air they breathe is severely polluted, leading to respiratory problems. The constant noise and burning light is such that they no longer know the difference between day and night.

Shell consistently ignores the Associated Gas Reinjection Decree, which the Nigerian government enacted in 1979, charging the oil companies to end the flaring of gas by 1984 at the latest.[52] Rather, the company prefers to pay the minuscule levy that the government later imposed as a penalty for flaring, refusing to take into consideration the havoc every cubic meter of flared gas is wreaking on the people of the Niger Delta and their already endangered environment. Even the flaring-reduction project that company officials had proposed for its Eastern Division operations, to collect about

25 percent of the gas flared and sell it to fertilizer and aluminium plants, was canceled in 1994 because Shell claimed there was no money to fund the project.[53] The Liquefied Natural Gas Project that Shell spin doctors now claim will lead to a "considerable" reduction in the amount of gas flared in the company's Niger Delta operations was not even designed originally to collect nonassociated gas, but had to be modified when the company's environmental policies came under fire in the wake of the Ogoni crisis. Nor has the Liquefied Natural Gas Project, which commenced operation in 1999, eliminated the incidence of gas flaring entirely.

In January 1999, Shell unfolded a new $8.5 billion scheme to expand its oil output in Nigeria by 600,000 barrels per day, claiming that the new project was part of a long-term strategy to commercialize the country's gas industry and also reduce the flares. Environmentalists, however, are worried that this increased output will come mainly from new offshore oil fields—the implication being that associated gas would continue to be flared in Shell's onshore oil fields. They also dismiss Shell's "vow" to eliminate flaring in its Niger Delta operations by 2008 as a ploy, arguing that the multinational, with its awesome financial and technical resources, could end all flaring in two years if it was really serious about doing so.[54]

Shell's production operations end at the oil terminals in Bonny and Forcados, where oil is collected in tanks and production water removed before it's loaded in tankers waiting offshore. Production water, already contaminated with oil, is discharged directly into the surrounding creeks and rivers without adequate treatment. Sludge and other lethal chemicals removed from the bottom of storage tanks in the course of maintenance activities are similarly disposed. Oil leaks from the storage tanks are also a regular phenomenon, and this, combined with evaporation directly from the tanks themselves, has subjected the soils, rivers, and creeks in the vicinity of the Bonny and Forcados oil terminals to slow but relentless devastation.[55]

Shell*ing* the Niger Delta to Death: Seven Case Studies

Shell in Ogoni

This Ogoni song, composed in 1970, sums up the Niger Delta community's experience with Shell:

The flames of Shell are flames of hell
We bask below their light
Nought for us serve the blight
Of cursed neglect and cursed Shell.[56]

The Ogoni, a rural closely knit community of farmers and fishermen, have lived in the central part of the Niger Delta for millennia. With a population of 500,000 inhabiting a total land area of 404 square miles, Ogoni is one of the most densely populated rural communities in the world. Yet, for all the population pressure, the inhabitants had over the years perfected the art of managing the resources of their environment in a sustainable way, so the land and the rivers and streams always yielded enough food for all. Indeed, Ogoni was once considered the food basket of Rivers State. The land was considered sacred, and to commit acts that polluted or desecrated it was viewed as an abomination and promptly visited with appropriate sanctions. The Ogoni saw themselves as custodians of the land and all that dwelled in it, including the rivers and streams, and went out of their way to protect and safeguard them all, knowing that their survival, indeed their very existence, depended on the well-being of their environment.

But all this was before the drilling rigs of Shell, in the words of the late Ogoni leader Ken Saro-Wiwa, "began to dig deep into the heart of Ogoni, tearing up farmlands and belching forth gas flames which pumped carbon monoxide and other dangerous gases into the lungs of my people."[57] Shell opened its first oil well in Ogoni near the village of Kegbara Dere in 1958. Before community protests at the oil company's destruction of their environment forced Shell to suspend its operations in 1993, SPDC had five big oil fields with ninety-six wells and five flow stations in Bomu, Korokoro, Yorla, Bodo West, and Ebubu. At their peak in the 1970s, the oil wells were producing an average of 108,000 barrels of oil a day. Indeed, Bomu was estimated at discovery in 1958 as a "giant" oil field, with at least 500 million barrels of oil reserves at the time.[58]

The Ogoni recognized the insidious impact of Shell's oil exploration and production activities on their environment from the onset. Letters written by Ogoni community leaders in April 1970, when the company was still a joint venture between Shell and British Petroleum, described how the multinational's activities were "seriously threatening the well-being, and even the very lives, of the Ogoni."[59] In yet another letter two months later, representatives of the Dere Students Association complained to Shell officials about the constant gas flaring and noise generated by the company's operations in the community. They also criticized the practice of laying oil pipelines over community farmlands and asked Shell to remedy the situation.[60] All these protests were ignored.

The first major incident of environmental pollution in the course of Shell's operations in Ogoni occurred shortly after the Nigerian civil war ended in 1970. There was an oil blowout in Well 11 in the Bori oil field. Ken Saro-Wiwa, who was then a commissioner in Rivers State, saw what happened: "I was witness to the great damage which the blowout occasioned to the town of Kegbara Dere. Water sources were poisoned, the air was polluted, farmland devastated. I watched with absolute dismay as indigent citizens found neither succour nor help from Shell."[61] For three weeks nonstop, oil spilled from the well and covered the surrounding creeks, rivulets, and rivers, leaving misery, widespread destruction, and disease in its wake. Lamented the people of Dere: "We no longer breathe the natural oxygen, rather we inhale lethal and ghastly gases. Our water can no longer be drunk unless one wants to test the effect of crude oil on the body. We no longer use vegetables, they are all polluted."[62] Shell, however, claims it was retreating Biafran soldiers that sabotaged the Agbada-Bomu oil trunkline in 1969 and caused the spill, a claim whose veracity was questioned by the Geneva-based World Council of Churches in its December 1996 report on the oil company's activities in Ogoniland.[63]

Some thirty years after the Kegbara Dere disaster, the community is still suffering the harsh aftereffects of the oil spillage. A three-acre portion of land in the community is still heavily polluted with crude oil. This oil evaporates into the atmosphere, thereby increasing its carbon dioxide and carbon monoxide content. During rainfall, these lethal gases dissolve in the water and descend as acid rain into the rivers and creeks and vessels that members of the community put out to collect drinking water. In a landmark judgment by Nigeria's Court of Appeal in 1995, Shell's claim that it had done enough to clean up the ecological damage in Kegbara Dere was dismissed out of hand. Justice J. C. A. Edozie's leading judgment put the economic and environmental devastation the community suffered as a result of the Shell well blowout in graphic perspective: "The blowout lasted for several weeks, during which time crude hydrocarbon, sulfur, and effluent toxic substances were violently emitted in dense fountains. The emissions formed a thick layer over the surface of the adjoining land, destroying farmland, crops, and economic trees and natural vegetation of the impacted areas with the resultant desertification of the impacted area of about 1,500 acres [belonging to several families] in K-Dere town."[64] Nnaemeka Achebe, a senior Shell official, was later to implicitly concede in a newspaper interview that his company had done virtually nothing to clean up its pollution in Ogoni because "only the Ogoni were making noise."[65]

On June 12, 1993, the same day presidential elections (the results of which were subsequently annulled) were held in Nigeria, a pipeline began to leak at the Shell flow station in Korokoro, Ogoni. For forty days the oil poured out, spreading into farmlands and water sources. Shell officials did nothing to arrest the situation, only to later turn around and claim that company engineers did not respond to the emergency promptly and repair the leaking pipeline because they feared they would be attacked by the community. The people of Korokoro vigorously denied this, pointing out that their distress call was deliberately ignored by Shell officials.

Whereas Shell takes great pains to bury its oil pipelines out of sight in other countries, in Ogoni the company simply lays them right across farmland and people's homes. Greenpeace commented on the company's pipeline network in the area in 1994: "Shell's high-pressure pipelines pass aboveground through villages and crisscross over land that was once used for agricultural purposes, rendering it economically useless. Many pipelines also pass within meters of Ogoni homes."[66] Shell officials claim that they adopted a policy of not burying pipelines in Ogoni because much of the area is swamp and this could increase the risk of fractures and oil leaks. They further claim that the company regularly reviews the positioning of pipelines and in fact reroutes them when they become hazardous, especially in areas where they impact on human habitation.[67] But the Ogoni dismiss this as cheap propaganda, explaining that their land is not swampy nor has Shell gone out of its way to reroute a single high-pressure pipeline it considers as hazardous.

The bulk of Shell's pipelines are rusty and obsolete, having been installed in the 1960s. In tacit admission of this fact, the company recently embarked on a belated pipe-replacement program. Still, Shell downplays the pollution figures in Ogoni and the Niger Delta in general and the impact on the environment. In 1992 the official line was that 60 percent of the oil spillage was a result of sabotage. Three years later Shell, put on the spot by environmentalists, was singing a new song. It admitted that 75 percent of spills in its operations in the Niger Delta was a result of old and corroded pipes, but still insisted that in Ogoni between 1985 and 1993, 69 percent of the spills were caused by sabotage.[68] Shell's new position has, however, not impressed environmentalists and the local communities and their spokesmen, least of all the late Professor Ake, who countered:"If Shell put out this, it can only be a smoke screen."[69]

Dante's Hell

The British environmentalist Nick Ashton-Jones visited the Niger Delta in 1993 and saw Shell in its true element, flaring gas indiscriminately into the atmosphere twenty-four hours a day. "Some children have never known a dark night," he reported.[70] From 1958, when it opened its first oil well in Ogoni, until community protests forced the company out of the area in January 1993, Shell had flared gas close to villages nonstop. As a fifteen-year-old secondary school student growing up in Bori, Ogoni, in 1958, Ken Saro-Wiwa, even at that tender age, recognized the adverse ecological impact of Shell's operations and began pseudonymously to write protest letters to the editor of one of the government-owned newspapers. In 1992, now spokesman of the Movement for the Survival of the Ogoni People (MOSOP), an organization he established in 1990 along with other Ogoni community leaders to fight against the ecological genocide perpetrated on his people by Shell, Saro-Wiwa attacked the company's gas-flaring practices directly at the Tenth Session of the Working Group on Indigenous Populations in Geneva on July 28:"The most notorious action of both companies [Shell and Chevron] has been the flaring of gas, sometimes in the middle of villages, as in Dere [Bomu oil field] or very close to human habitation as in the Yorla and Korokoro oil fields in Ogoni. This action has destroyed wildlife and plant life, poisoned the atmosphere, and therefore the inhabitants in the surrounding areas, and made the residents half deaf and prone to respiratory diseases. Whenever it rains in Ogoni, all we have is acid rain, which further poisons watercourses, streams, creeks, and agricultural land."[71]

In response to this criticism, Shell has publicly stated that its flares are usually located away from human habitation and adequately protected by earth bunds. Company officials argue that there is population pressure in Ogoni and blame the local communities for building their homes close to the gas flares in their desperate scramble for new land. Said Shell officials, "When communities have expanded in the direction of production facilities, SPDC has taken appropriate action, including relocation of flares." They also claim there is no evidence that flares affect crops.[72]

But Shell has not been able to produce clear evidence to show that it actually relocates gas flares away from populous areas in Ogoniland. Nor is its claim that gas flares do not affect food crops supported by science. Van Dessel, a biologist and former head of Environmental Studies at Shell Nigeria, has argued that because combustion temperature and efficiency at Shell flow stations are relatively low, "this has important consequences for the contribution of flaring to the greenhouse effect."[73] Project Underground led a group of local and international observers to Luawi, an Ogoni village, in April 1997. The team analyzed samples taken from the village stream and discovered that total petroleum hydrocarbons tested at eighteen parts per million (ppm), 360 times higher than levels deemed acceptable in the European Community.[74] Those who wear the shoe know where it pinches, and what better rebuttal of Shell's dubious "no harm to crops" thesis than the testament of the local communities themselves: "Apart from physical destruction to plants around flaring areas, thick soots are deposited on building roofs of neighboring villages. Whenever it rains, the soot is washed off and the black inklike water running down the roofs is believed to contain chemicals which adversely affect the fertility of the soil."[75]

Shell in Nembe

Shell has five flow stations in Nembe in the southern Delta. On March 8, 1994, a major oil blowout occurred in the Nembe 2 flow station, which has twenty-four wells connected to it. Incidentally, a journalist from the respected Nigerian daily *The Guardian* was on hand to witness the environmental carnage:

> The approach to Nembe 2 has a telltale smell about it. There was evidence of heavy oil pollution in the air. The surrounding blocks of mangrove forest looked vacant, deathly, and partially scorched. The nearer we got to the rig, the hotter the air and the stronger the smell of crude

> oil. Soon the entire surface of the creek was coated with a thick layer of glistening-brown crude oil . . . Nembe 2 was a disaster this afternoon. It has about 24 wells connected to it, and the blowout could have occurred at any one of them.[76]

Blowouts and leaks from Shell pipelines and well heads, as well as constant gas flaring, have become a fact of life for this riverine community, disrupting the normal rhythms of daily life and threatening their economic survival. Community leaders say an average of fifteen oil spills occur each year. Between December 10, 1993, and February 1994 they counted eleven spills, and they complain that the magnitude was such that it took the town over two months to resume normal social and economic activity. They also blame Shell officials for not reacting promptly when these spills occur, and cite as an example a spill from a major distribution line that happened in February 1994 and lasted for three days, despite the fact that members of the community notified Shell as soon as they noticed the incident. Shell had no adequate rapid response mechanism in place to deal with such emergencies, and although its officials worked day and night to stem the spill, it had caused considerable damage to the vicinity by the time the faulty pipeline was eventually capped.

The majority of Nembe people are fishermen, and they have borne the brunt of the spills. The first victims are the fishing nets in the creeks and streams. As many as a thousand can be damaged in a single spill. The fish ponds and lakes also receive a lethal dose of oil, which effectively expels or destroys the fish population. Tidal waves then help to spread the oil quickly through the mangroves, destroying forests and soil in their wake. The Nembe ecosystem is direly threatened, and there is a sense in which it can be said that the community is dying in installments. Some of the palm trees in the community no longer bear fruit as a result of oil-polluted soils. Local people claim that some species of tilapia in the creeks, in the vicinity of gas flares, have lost the ability to change color and become rubbery when cooked. The people of Nembe lament that certain staples no longer thrive on Nembe soil. The red and white varieties of cocoyam and the mammy yam known among the locals as "Mercy" have become extinct. Local varieties of rice have also been recording declining yields.

Obasi Ogbonnaya of *The Guardian* paid a visit to Nembe town itself after witnessing the blowout in Nembe 2 flow station, and reported that due to the lingering effects of recurrent oil spills, Nembe fishermen now spend eight hours and more out on the river at long distances, only to return with no catch at all. Indeed, the delicacy in Nembe today is frozen fish bought in the city and ferried in at exorbitant costs.[77]

Shell in Okoroba

Okoroba, a small pastoral village near Nembe, was an ecologically balanced, self-sustaining community before Shell's earthmovers came calling in 1991. The environment is not only protected but also revered in Okoroba, with a long catalogue of customary laws put in place to ensure that the land, rivers, and creeks, and all that thrive in them, are not unduly harmed or destroyed. Such is the seriousness with which the people take their role as custodians of their environment that first-time visitors are warned right from the outset: "You are not allowed to touch any blade of grass or leaf of any kind or kill any animal here."[78]

This harmony was violently disrupted in 1991 when Shell discovered oil in a site a third of a mile from the village. Without the benefit of an Environmental Impact Assessment—at least, none was shown to the local people—Shell officials decided to construct a two-and-a-half-mile canal (slot) right across the village in order to transport heavy drilling equipment, which would be used to construct an oil well head at the site. Okoroba village is delicately sandwiched between a freshwater swamp, which provides the community with drinking and cooking water, and a saltwater area where periwinkles and other seafoods are sustainably farmed using the traditional shifting cultivation method.

Even though the well site could easily be reached over land, Shell chose to construct the canal to enable it to float in heavy drilling equipment, causing the freshwater and saltwater areas to mix together, thus triggering an ecological disaster in Okoroba.[79] Environmentalists who visited the village to assess the impact of Shell's oil exploration activities in the community reported:

> [Shell] directed that several coconut plantations along the river be destroyed and the old semicompleted hospital bulldozed. The dredgers destroyed the Ighobia creek, dug and dumped mud by the side of the river, uprooted graves, pumped water into farmlands, forests, ponds,

> and lakes. Shell contractors also killed sacred animals, desecrated shrines, and scared many wildlife away to vulnerable areas where they were subsequently slaughtered or shot for "bushmeat."[80]

Shell's activities in Okoroba and environs, particularly the dredging up of mud, has contributed to the gradual silting up of the Okoroba River, disrupting its flushing and regulatory dynamics. Whole families, numbering about six thousand people in all, lost their food and cash crops after Shell's dredgers had passed through the community. The people of Okoroba who depended on the river and the outlying farmlands for their livelihood suddenly found themselves trapped in a ravaged environment that could no longer provide them with succor. And as the drilling of the oil commenced, a lot of oil was allowed to flow into the river, polluting it and further compounding their woes.

The people of Okoroba are still counting their losses. It has been estimated that at least twelve people have died in the aftermath of Shell's ecological war in Okoroba. A BBC journalist, Alice Martin, visited the area in early 1995 and wrote this report, based on her personal observations and detailed interviews she conducted with community leaders and ordinary citizens in Okoroba:

> The story of the devastation of villagers' lands and livelihood is terrible. The fishes have died in the creeks and rivers because of the mixing of fresh- and saltwater, and it takes fishermen two to three days to "pull out" to good fishing grounds now. Even children no longer catch "small fry" for their mothers because of the danger of strong currents and pollution of the water. Crops have died and great stretches have become infertile due to water-logging caused by the high muddy banks along the sides of the channel. Above all, the water is polluted, "sometimes purely green, sometimes purely blue" . . .[81]

The destruction of farmlands and fishponds by Shell's canal in Okoroba has radically altered the economic life of this once self-reliant community—for the worse. Obebara Douglas, an aged farmer, lost his banana and coconut fields and seven fishponds to the canal, which has caved in in places. Several orange orchards also had to give way, and the people of the community have expressed fears that in addition to the dying fish population, the polluted soils can no longer support cassava, a favorite local staple. The cost of living has soared in Okoroba as a result. Where once fish, an important part of daily nutrition in these parts, was obtained virtually for free in the fertile fishponds, it is now sold at mind-boggling prices.[82]

A major spill occurred in Okoroba in mid-1994 and laid waste to huge swaths of vegetation.[83] Village hunters complain that such animals as the bush pig, crocodile, iguana, and the monkey, upon which their livelihood depend, have either been wiped out or forced to flee the polluted forests. Rare animal species like Sclater's guenon (the long-tailed monkey, and the only mammal species known to be endemic to Nigeria) and the chimpanzee have also fled from designated safe areas that the village's official hunters set up specifically to protect them as a result of dying undergrowth caused by the oil spill from Shell's leaking well head.

Okoroba's efforts at economic and ecological sustainability are now severely threatened. The villagers fear they will have to move to another location or face extinction in a land so mangled by Shell that it can no longer support them as it once supported their ancestors. The process has already begun. So contemptuous was Shell of the people's customs and way of life that its dredgers simply dug up ancestral graves and dumped the remains of Okoroba's dead into the ocean. Lamented Chief Famous Alawari of Okoroba, "We don't bury and go to remove the bodies. That is not our custom. But forcefully they [Shell] removed the graves to the ocean to get access roads to their well head."[84] In Okoroba, even the dead are not allowed to rest in peace.

Shell in Iko

Iko is a small rural village on one of the sand barrier islands in the Niger Delta.[85] Iko Community, of which this village is part, has an estimated population of 100,000 and belongs to the Ijo ethnic group in Akwa Ibom State. In 1974, Shell opened its Utapete flow station in Iko village. By 1982 the company was extracting nearly 10,000 barrels of crude oil per day from Utapete.

Shell's operation has brought misery and environmental destruction to Iko instead of the wealth normally associated with crude oil. An Environmental

Rights Action team, following a desperate plea from members of the Iko community for assistance in their struggle against Shell and its environmentally hostile policies, visited the village between February and June 1995. The ERA team conducted an in-depth field survey to determine the nature and extent of the impact of the oil company's operations on the community and their environment, supplemented with the reports of government officials who had also carried out their own studies in Iko independent of ERA.

The result of the various surveys, authenticated by the Akwa Ibom State Environmental Protection Agency (AKSEPA), a government agency, is as follows:

a. There is no flaring stack at the gas flaring site at Shell's Utapete flow station, located less than 250 meters from the nearest residential building in Iko. This negates the provisions of the international code of practice in oil fields which Nigeria acceded to as a member of OPEC. Shell stopped the flaring following ERA's exposure of this practice in July 1995.
b. There was visible destruction of the mangrove vegetation, probably arising from the effects of oil pollution along Iko beach. The banks and creeks draining the flow station at Utapete are stained with crude oil.
c. Most buildings in the community, especially those with corrugated iron sheet roofs, experience massive corrosion damage resulting in frequent changes and leakage. The people have now resorted to the use of thatch or mat on top of the CIS roofing sheets to minimize the damaging effects.
d. The black smoke emitted into the atmosphere from flares is indicative of pollution of the atmosphere chemically and thermally over a long period, considering that the flow station emits gas twenty-four hours a day nonstop. (This practice has been stopped.)

Edem Esara, an environmentalist in the employ of AKSEPA, also conducted an extensive study of Iko as part of his graduate degree research and observed a high concentration of sulfur in all the environmental data he analyzed. Said Esara: "This effectively indicts the gas flaring activity of the company [Shell], which is the only possible source of this pollutant within the study area. The value obtained in the study (sufficiently corroborates the claim by a Shell staff) to the effect that most of the oil wells feeding the Utapete flow station are very rich in sulfur compounds."[86]

The result of the above scenario is acid rain, which the people of Iko say has been corroding the zinc and nonzinc roofing of their houses and also harming plant and animal life. Complained Sunday Ikare, a community leader, "The sulfur combines with salt to eat the roofs off the houses of our people."[87] Studies have shown that the ability of sulfuric acid, produced when SO_2 reacts with atmospheric water vapor, to damage soil, buildings, and other structures is especially obvious where large amounts of sulfur-bearing oils are flared. There is also evidence that sulfuric acid mists cause damage to plant leaves and that sulfur particles, combined with other lethal gases in the atmosphere, have been responsible for extensive damage to forests in the areas near large combustion sources.[88] The adverse impact of acid rain on rivers and surface waters is no less significant. Fish and, in particular, their eggs and juvenile forms are especially susceptible to increases in acidity, decreasing in population or dying off completely.

Shell officials have argued that Nigerian crude is generally low in sulfur content and so there can be no possibility of acid rain in Iko as a result of gas flaring. The company further claimed that it commissioned an independent study by consultants from the University of Calabar (in southeastern Nigeria) and they found that acid rain was not widespread in the area. Significantly, the consultants established that acid rain fell in the Niger Delta for one month in the year but claimed they found no relationship between this occurrence and gas flaring.[89] Edem Esara, however, pointed out that the sulfur content of crude extracted from five of the eight Shell oil wells in the area is particularly high, and that given the huge amount of gas and oil that is flared daily in relation to the small size of the community, the Iko atmosphere is literally suffused with sulfur particles, which eventually descend as acid rain.[90]

Iko has also suffered several incidents of oil spillage since 1973. In October 1994 a big blowout occurred at the Utapete flow station. Six villages, including Iko, were affected. Fishes died and water sources were polluted. When members of the community attempted to take photographs of the carnage, policemen forcibly prevented them from doing so. It was thus difficult to ascertain the exact amount of oil spilled, but so extensive was the damage that Shell officials had to shut down the flow station for two weeks.[91]

In 1994 and again in 1995, Shell laid the blame for blowouts in Iko on other oil companies operating in the area. On these two occasions, Shell claimed that the oil spillage came from Mobil installations offshore. However, an AKSEPA official who investigated the incident told ERA: "We have confirmed that it is Shell. No matter their denial, we have our facts. The oil conglomerate should stop the game of deceit and own up to the double standards it is practicing in its oil operations in Nigeria."[92]

Gas flaring and oil spillage from Shell's Utapete flow station have taken a terrible toll on the Iko environment and the people themselves. Crop yields have been declining, as well as the fish population in the Iko River and the surrounding creeks. Life expectancy is on the decline, and the people complain of respiratory ailments, which they attribute to the polluted atmosphere. So outraged was the ERA team that it recommended that the Utapete flow station be shut immediately. The report concluded:

> Arising from the improper environmental management and inadequate moral, financial, and technical investments by Shell in Utapete flow station, and because of the high likelihood of health hazards through air, soil, and water pollution, the effects of which are already devastating on the physical, biological, and cultural components, the Shell flare should be shut down permanently.[93]

Shell in Bonny, Forcados, and Ughelli

Shell processes about 595,000 barrels of water a day along with oil and gas from its various fields in the Niger Delta.[94] This is water pumped into oil wells to facilitate extraction. It is brought up with the oil and then sent through pipelines to the two storage terminals in Bonny and Forcados, where the oil is separated and loaded on offshore tankers. The company also operates a Quality Control Centre at Ughelli (UQCC), where 150,000 barrels of water are processed daily. Bonny and Forcados handle oil produced in Shell's Western Division, while UQCC serves the Eastern Division.[95] After the oil has been removed and sent to the tankers, the water is dumped into the surrounding swamps, creeks, and rivers. Shell claims this water is treated to meet the legal limit for oil and grease in effluent waters.

In the course of separating the oil from water, Shell officials use chemicals to induce settlement in the tanks. The end product of this separation process is thick, oily sludge which combines with firefighting chemicals like Halon, already in the tank, to form a potent mixture.[96] This hazardous substance is then discharged into the swamps and rivers. The Bonny River estuary, the swamps around Forcados, and the Warri River near Ughelli, where Shell discharges its production water, have all been contaminated after nearly four decades of receiving this cocktail of dissolved and dispersed hydrocarbons, sludge, and firefighting agents.

The World Bank has explained that the oil separation process generally results in the release of less than one-third as much oil into the environment as what is caused by spills. Using figures supplied by Shell, bank officials estimate that an average of 710 tons of oil is discharged by the oil companies in the Niger Delta annually.[97] However, as Greenpeace has argued in respect to Shell's terminals, "very large quantities of water are involved, [and] the final quantity of oil thus discharged into the environment is considerable."[98] Greenpeace also disputes Shell's claim that it discharged an average of 335 tons of oil annually in the course of its separation process, pointing out that various studies indicated that there were much higher concentrations of oil in discharged production water.

In the briefing notes it hands out to journalists and environmental groups, Shell claims that its production water is first treated with a number of gravitational techniques and is monitored regularly before being discharged into the Bonny River estuary. Company officials also claim that the hydrocarbon content of its production water averages 7.8 milligrams per liter and that this does not constitute an environmental hazard. Yet a 1993 environmental impact study commissioned by Shell itself found an average hydrocarbon content of 62.7 milligrams per liter in Oloma Creek, near the Bonny oil terminal.[99] World Bank officials took a look at these figures and declared that they were unacceptably high and indicate poor or no treatment of effluents by Shell in its oil terminals. The World Bank also criticized the API separators and TPF basin facilities that Shell uses to separate the oil from water, saying that the method can achieve a standard of only 50 milligrams per liter at best, far short of the universally acceptable maximum of 20 milligrams per liter.[100]

The Bonny terminal was built in 1961, and like Shell's rusty pipelines, has not been replaced.[101] Shell claims it has recently commissioned a project to upgrade the Forcados terminal and that when this is completed, Forcados would take over all water disposal in the Western Division. There is, however, no talk of commissioning a major study to determine the ecological impact of the production water it has discharged into the water systems of the three communities of Bonny, Forcados, and Ughelli these past decades. According to Professor Ake, "The mud at the bottom of the Bonny River has a lethal hydrocarbon concentration of 12,000 ppm."[102] This spells death for flora and fauna, and sickness for the local inhabitants who depend on them for sustenance.

It is expected that the volume of production water processed by the oil companies will have doubled by the year 2000 as pressure inevitably decreases in old oil fields, thereby requiring more water injection to bring up the oil.[103] This, in turn, is expected to double the rate of environmental pollution. Given this scenario, the World Bank fears that it will be difficult if not impossible to rehabilitate the human ecosystem of Bonny and all the other communities that Shell, in concert with the other oil companies, has subjected to severe degradation.[104]

Shell and the Liquefied Natural Gas Project

Shell has found in the Liquefied Natural Gas Project (LNG) a perfect handle for its claim that it is now taking corrective measures to limit the adverse effects of its operations on the Niger Delta environment. According to the company's press releases, the LNG will lead to a "considerable" reduction in the amount of associated gas flared into the atmosphere. What company spin doctors do not divulge to the public, though, is that the original design of the $3.8 billion project did not include the utilization of associated gas and had to be hastily built in following international pressure on Shell to put a stop to its environmental devastation of the Niger Delta. Even at that, the amount of gas flared by Shell will be reduced by only 20 percent.[105]

The LNG Project had actually been on the drawing board since 1964. After several false starts, the final papers were drawn up and signed in December 1995, establishing Nigeria LNG Ltd., a joint venture between NNPC (49 percent), Shell (26.5 percent), Elf (15 percent), and Agip (10.4 percent). Shell is the operating company.[106] The International Finance Corporation, a World Bank affiliate originally expected to take up 2 percent of the project, pulled out in the wake of the execution of Ken Saro-Wiwa in November 1995, explaining that the political situation in Nigeria was such that it could no longer invest in the country.[107] The LNG Project is designed to collect associated and nonassociated gas from gas treatment stations operated by SPDC, Agip, and Elf, and cool it to minus 162 degrees Celsius, so that it becomes liquid, taking up one six-hundredth of its previous volume and thereby making it cheaper to transport. This liquid gas, processed in a new gas plant in Bonny, is then transferred to four tankers for sale in European and American markets.

The LNG Project began to produce 6,600 tons of liquefied natural gas daily in November 1999. SPDC supplies 53.3 percent of the gas drawn from its Soku, Nembe Creek, and Ekulama oil fields.[108] Agip and Elf supply 23.3 percent each. The site for the gas plant was cleared of vegetation in 1979, and the people of Finima village who originally occupied it were relocated in 1991-92.[109]

The Nigeria Liquefied Natural Gas Project consists of two main elements: the LNG plant and its associated facilities on Bonny Island, and the Gas Transmission System (GTS), a network of pipelines that snakes through mangrove, freshwater swamp forest, dry land rain forest, farmland, freshwater rivers, and brackish and estuarine areas. Environmentalists, before the project went on-stream, were worried about its potential ecological impact on the already endangered Niger Delta ecosystem and said that neither Shell nor the Nigerian government, the two principal partners in the project, had done enough to ensure that these potential harmful effects were minimized.[110]

Separate environmental impact assessments were conducted for the LNG and GTS, the first in 1989. A second one was carried out by the Liverpool-based SGS Environment Ltd. in 1995. The thrust of its report confirmed the fears already expressed by such environmentalists as Nick Ashton-Jones, that an improperly designed LNG Project, and one whose operations were not monitored strictly to ensure conformity with international environmental standards, could turn out to be an ecological nightmare and an economic disbenefit for the local communities. Outlining the potential impact of the project, the SGS report said: "The main impact of the GTS construction on the natural environment is the loss of ecologically important habitat and the fragmentation of what remains. There will be substantial impact on mangrove and freshwater swamp forests and their soils and groundwaters, which cannot be mitigated in full."[111]

The report then went on to catalogue seventeen other expected environmental effects of the project, among which were a sudden increase in the population of the communities in and around Bonny, with the social and economic pressures this will inevitably generate; the introduction of such vector-transmitted diseases as malaria, due to ponding arising from construction activities; and general degradation of the environment and loss of biodiversity in the mangroves and freshwater swamps in the area, in particular due to increased human activity. The SGS report also outlined ten preventive/repressive mitigations and four curative/compensation measures it hoped would ensure that the adverse environmental side effects of the LNG Project are reduced to an acceptable minimum, ending on the hopeful note that "Nigeria LNG Ltd. recognize that the development, implementation, and maintenance of a system for the long-term management of the environment must form an integral part of business quality management. To this end, they have developed an Environmental Management System (EMS) to achieve these aims."[112]

While the potential economic benefits of the LNG are not in doubt, given that it will boost Nigeria's foreign exchange earnings while at the same time helping to reduce the flaring of associated gas, critics of the project have pointed out clear lapses in the Shell-commissioned EIA and say that not only did it not address the problems on the ground squarely, its general contents fell far short of international standards. In a letter to the International Finance Corporation in July 1995, the British environmentalist Nick Ashton-Jones highlighted some of the shortcomings of the SGS report:

> The worrying thing about the SGS environmental assessment is that it assumes that Nigeria environmental legislation will work: by and large it does not, although it is an invaluable handle for NGOs to use. The report on the assessment states: "Development planning control in Nigeria is currently based on the Fourth National Development Plan (1981–1985). The FEPA [Federal Environmental Protection Agency] was established under the Environmental Protection Act, under which the development and monitoring of environmental standards was intro-

> duced. Environmental Impact Assessment became mandatory for major projects in 1992." This really is a fairy tale, and any sensitive team of environmental impact assessors working in Nigeria should have picked up the fact: it really does not work, and there is no guarantee that recommended mitigation of the adverse environmental impacts of the LNG pipeline will be implemented and maintained in the long run. The EIA has ignored the political and social realities of Nigeria . . . Also, reading the report, it is difficult to believe that anyone involved actually traveled the entire length of the pipeline; for instance, no places were mentioned in Ogoniland. In a free and open society I do not think that the report would be acceptable to a public that was allowed to express its opinions freely.[113]

Shell's involvement in the LNG Project as operating company has also been cause for serious alarm, given the company's poor record in environmental protection and respect for human rights in Nigeria. The argument of concerned environmentalists is that there is no reason to expect Shell to behave any differently in its management of the gas project. This view finds support in yet another report, commissioned by The Body Shop International in December 1995 to review the SGS Environmental Statements. The review, while conceding that there are some good points about the SGS reports, nevertheless criticizes their theoretical and abstract approach:

> The reports do not properly address the hazards and risks associated with an LNG facility or the gas pipelines. A full hazard and risk assessment should have been carried out on the pipelines, LNG plant, and transport of the LNG by the six cryogenic vessels. This assessment would need to take into account the political instability in Nigeria and the fact that during the life of the facility (perhaps forty years) Nigeria could become so inhospitable for foreign companies that they are forced to withdraw their expatriate personnel. The problems associated with sabotage of LNG facilities should also have been defined. LNG has the ability to spread very quickly through underground drains and sewers until it finds an ignition source, when it ignites the whole vapor cloud. This does not appear to be mentioned in the reports.[114]

In other words, the Shell-commissioned EIA was deemed to be neither adequate nor comprehensive, a reflection of the company's attitude toward the environment and the local communities that are listed in its press releases as a major beneficiary of the LNG Project. Also disturbing is the fact that Shell has not seen the need to conduct a postimpact assessment in New Finima following the relocation of the people of the village. Employment in New Finima is a paltry 26 percent, compared to 72 percent in Bonny town.[115] Why is this? Significantly, the SGS report acknowledged that the LNG Project would create considerable social and economic problems in Bonny and the surrounding communities, but added glibly that these problems would be smoothed away with "efficient public relations."[116]

"Efficient public relations" proved useless when youths in Bonny town, the main operational base of the gas project, stormed the LNG facilities in September 1999, blockading the gas plant.[117] Community leaders had warned LNG officials in July that youths in the town were becoming restive following the nonimplementation of the provisions of the Memorandum of Understanding reached by both parties, providing for jobs for local people and also a comprehensive development program for Bonny town.[118] Community leaders also accused company officials of failing to consult the Bonny Kingdom Development Committee on key developments affecting the town, of not employing qualified residents in the project, and also ignoring their request that they be represented in a new environmental committee established to oversee the operations of the gas project.

After waiting for several weeks to no avail, angry youths took over the access road between the main gas plant and staff residential quarters on September 25. During the scuffle that ensued, an expatriate staff member of the project shot and killed several of the protesters.[119] It was not until President Olusegun Obasanjo flew into the troubled town in a helicopter two days later and sued for peace that the youths lifted the blockade. At a meeting with President Obasanjo, the youths and elders of Bonny complained that the LNG officials had failed to adhere to the mitigating measures spelled out in the Shell EIA report, and demanded immediate employment for local youths in the gas project, a penalty to be imposed on LNG for gas flaring, which was still a recurrent feature in

Bonny, and the repatriation and punishment of the official who shot some protesters.[120]

The Open Sore of a Fragile Ecosystem

All over the Niger Delta a terrible tragedy, more chilling for its fragmentation into just another day's sad story in a thousand and one small communities, is quietly playing itself out. One prominent landmark in Otor-Udu, a small community in the western Delta, is Shell's Utor Well 17, which was drilled right into the heart of the village. Children, oblivious of the health hazards, have converted the drilling waste pits near the well into a swimming pool. Day and night the ever-present threat of oil spillage and all that it entails looms large over the doomed village like a forbidding cloud. Humphrey Bekaren, a journalist born in Otor-Udu, has written about the Shell operations' devastation of his village and the other communities that comprise Udu Kingdom:"Every village is a witness to oil misfortune. If there is no oil or gas well, then there will be the open-mouthed burrow pits used for sand-filling roads and locations, or the ubiquitous crude oil pipelines which the oil companies prefer riding unmindfully above the ground in a dense network, thereby blocking off farmlands and water. The gas flares turn night to day and days to hell; while the heavy-duty trucks rumble through the land."[121]

In the Ogbia Local Government Area of Rivers State, freshwater supplies for a number of communities were disrupted by Shell's engineering works in 1993, resulting in a serious outbreak of cholera. Shelley Braithwaite, an Australian environmentalist who conducted an environmental and social investigation into Shell's Nigeria operations in 1998, examined drinking water samples from five sites in the Niger Delta where Shell has installations—Otuogidi, Aleibiri, Oruduba Creek, Biseni, and Ihuowo—for total petroleum hydrocarbon (TPH) contents. She discovered that TPH in drinking water in all five communities ranged from 250 to 37,500 times the legislated level of 0.01 ppm for untreated drinking water within the European Union.[122] Wrote Braithwaite, "Social investigations in the rural communities of Aleibiri, Biseni, Ihuowo, and Otuogidi revealed widespread discontent toward Shell. Factors contributing to the complaints, while site specific, display a regional accordance and lie with the environmental and socioeconomic impacts of oil spills, the protracted and substandard quality of clean-up operations, and the lack of adequate compensation."[123]

In June 1999, Michael Fleshman, human rights coordinator of the New York-based The Africa Fund, also visited the Ijaw village of Otuegwe 1, where a sixteen-inch underground Shell pipeline had burst in June 1998, discharging an estimated 800,000 barrels of oil into the surrounding creeks. The trip was a defining moment for Fleshman. He reported:

> The impact of the spill on the community has been devastating, as the oil has poisoned their water supply and fishing ponds, and is steadily killing the raffia palms that are the community's economic mainstay. Lacking any other alternative, the people of the village have been forced to drink polluted water for over one year, and the community leaders told us that many people had become ill in recent months and that some had died. The sight that greeted us when we finally arrived at the spill was horrendous. A thick brownish film of crude oil stained the entire area, collecting in clumps along the shoreline and covering the surface of the still water. The humid air was thick with oil fumes.[124]

Irri, Yenogoa, Diobu, Peramabiri, Kiolo, Odioma, and several other communities—all have, like Otuegwe 1, tasted what it is like to have Shell as a next-door neighbor, and it is not an experience they will recommend to other communities anywhere in the world.[125] Slowly, quietly, the Niger Delta is being mauled beyond repair.

Environmental Rights Action has always argued that all ecosystems are, in the final analysis, human ecosystems, and that the human ecological relationship between local people and mining companies such as Shell must be in harmony if the Niger Delta is to return to the path of sustainable development. By its deeds these past forty years, however, Shell does not appear to think so. Ken Saro-Wiwa, before he was murdered by the Nigerian junta in November 1995, had accused the oil company of waging an ecological war against his people. Said Saro-Wiwa, "Thirty-five years of reckless oil exploration by multinational oil companies has left the Ogoni environment completely devastated. Four gas flares burning for twenty-four hours a day over thirty-five years in very close proximity to human habitation; over one hundred oil wells in village backyards; and a petrochemical complex, two oil refineries, a fertilizer plant, and oil pipelines crisscrossing the landscape aboveground have spelled death for human beings, flora, and fauna. It is unacceptable."[126]

For Ogoni read the "Niger Delta." And the Niger Delta is the world.

FIVE

Where Vultures Feast

[Shell] made a lot of promises: the hospital and toilet houses were destroyed, as were the burying grounds. They pumped out water and destroyed the farmland with promised compensation like community and secondary schools, a road to Nembe and pipe-borne water. But they left and nothing has happened. It is like dreamland.

Okoroba community leaders in November 1993

The Road to Oloibiri

Oloibiri is usually depicted in Nigerian schoolbooks as a scenic pastoral village where oil was first struck in commercial quantities by Shell in June 1956. All you see is lush green mangrove, picturesque little houses, and children playing happily in the square without a care in the world. But there is nothing romantic or beautiful about the real Oloibiri. Said a British Petroleum engineer of the village in 1990, "I have explored for oil in Venezuela, I have explored for oil in Kuwait, I have never seen an oil-rich town as completely impoverished as Oloibiri."[1]

Edwin Ofonih, fifty, is the government tax collector in Oloibiri. He is also a native. Ofonih still remembers with nostalgia the day Shell engineers discovered oil in the village: "In 1956, I was with my father when crude oil was found at eleven o'clock in the morning. We thought that when oil was found—we people out here are very poor—we thought we would be millionaires. We are still depressed. The town is very tattered. Shell promised to build schools and to make a sea wall because the town is flooded every year. Nothing was done."[2] Forty years after Shell lugged its drilling rig into the village, extracted all the oil, and left, Ofonih and the eight thousand other residents of Oloibiri are still waiting for the promised riches.

There is no proper road linking Oloibiri to the outside world. Travelers have to navigate their way through treacherous creeks and rain forest. There is no drinkable water in the village. The government hospital begun in 1972 has since gone the way of all abandoned projects, inhabited by rats and cockroaches instead of patients needing care. Said Chris McGreal, a British journalist who visited Oloibiri in December 1995 to find out what life was like in the village that once poured millions of dollars into the coffers of Shell and the Nigerian government: "What little has come to the town has been through the work of its residents or by the grace of the Rivers State government. [Electricity mains were] not installed until last year [1994]. Shell's sole contribution is a six-classroom extension to the secondary school."[3]

The general feeling in all the oil-producing communities in the Niger Delta where Shell has its operations is that the company has subjected them to gross exploitation these past four decades, taking away the oil from their land and giving them virtually nothing in return. Shell's standard response, when confronted with evidence of widespread poverty and neglect in an area that accounts for 13 percent of its worldwide profits, is that 90 percent of the oil revenue from the Niger Delta goes to the Nigerian government, that it shares just a dollar per barrel with Elf and Agip, the two other partners in the joint venture, and that "it is not for Shell to say how its contribution to the national purse should be spent."[4] Company officials also say it is not the responsibility of Shell, a foreign business venture, to develop the oil-producing communities, and that the federal government ought to be taken to task on the question of poverty and neglect in the Niger Delta. Said Chidozie Okonkwo, community and environment manager in the company's Western Division, "It is not really the business of a private company to develop [these areas]. So, when we get involved in infrastructural development, it is a complement to government's efforts."[5]

Implicit in this argument is that Shell is just another company doing business in Nigeria, handing over the bulk of the oil revenue to the government and keeping just a tiny fraction for itself, out of which it goes out of its way to fund development projects in its client communities. However, community leaders in the Niger Delta, and indeed other watchers of the Nigerian oil industry, advance the counterargument that Shell is not just "another" company in Nigeria, given its preeminent role as the country's foremost income earner. They say that while they accept that the company is not legally required to help develop the oil-producing areas, it is nevertheless morally obliged to plow back a fair portion of its profits, running into millions of dollars annually, toward the social and economic development of the communities where this great wealth is generated.

Shell officials habitually reel off figures and statistics, claiming that contrary to the loud complaints of neglect in the Niger Delta, the company has in fact been contributing more than its fair share to the development of the area since it struck oil in Oloibiri in 1956, and that it established a community assistance program in its areas of operation over twenty-five years ago. In a pamphlet titled "Community 1996" and issued in Lagos, Shell officials claimed that the company spent $36 million on its community program in 1996.[6] In a briefing note distributed to journalists in the wake of Ken Saro-Wiwa's murder in November 1995, Shell also claimed that "the company has stepped up support for the communities, in recognition of the lack of development, and it is now spending some $20 million a year in its area of operations on community projects."[7] These figures, however, are supplied by the company's public relations department and have not been independently verified.

Partners in Progress or Just Plain Parasites?

Perhaps a more fruitful way of assessing the validity of the company's claim that Shell and the communities are really partners in progress and that it is doing all it can for them in the way of development assistance projects is to pose the simple question: Exactly how much is its oil concessions in the Niger Delta worth to Shell annually, and what percentage of this does it spend on the so-called community assistance programs? Restating some of the known facts: Shell controls about half of all the oil concessions in Nigeria. Of the two million barrels of oil produced by the country every day, some 800,000 to one million barrels is produced by Shell.[8] Globally, Nigerian oil accounts for almost 14 percent of the company's production, the highest outside the United States.[9] Although Shell is engaged in oil and gas production in forty-five countries, Nigeria alone generates 13 percent of its total profits annually.[10] So exactly how much is Shell Nigeria worth?

According to the company's own figures, Shell, along with the other two joint venture partners, earns a net profit of one dollar on every barrel of oil it produces in Nigeria. Of this amount, Shell takes two-thirds as its own profit and gives the rest to Elf and Agip.[11] Based on these figures, Greenpeace estimated that Shell earns between $530,000 and $670,000 a day, or $200 million every year.[12] These, however, are rough statistics and tend to present the company as doing all the hard work while Nigeria's indolent and corrupt government takes the lion's share of the results.

The Memorandum of Understanding (MOU) that the military regime of General Ibrahim Babangida signed with the oil companies in January 1986, subsequently revised in 1991, offers a better clue to Shell's real earnings from its Nigeria concessions.[13] Going by the so-called gentlemen's agreement between the government and the oil companies, the companies are entitled to a guaranteed profit of between $2 and $2.50 per barrel produced as long as oil prices remain in the $12.50 to $23.50 bracket, and provided they can show evidence that they invest a minimum of $1.50 on every barrel they produce.[14] As a further sweetener, Shell and the other oil companies are entitled to an extra bonus of ten to fifty cents per barrel for every operational year they discover new oil fields with reserves greater than the volume of oil they extracted.[15] Shell has been laughing all the way to the bank since the MOU became operational. On average, world oil prices have not plummeted below the stipulated $12.50 lower margin since 1986.[16] This means that the company, along with Elf and Agip, has been earning a minimum of $2.30 on every barrel of oil it produces daily. While it is true that a percentage of this profit is plowed back into further exploration, it is for the future good of the company.

In 1993, SNEPCO, a wholly owned Shell subsidiary that has been prospecting for oil in the environmentally sensitive Gongola Basin in northern Nigeria, signed a production-sharing contract with the government. The terms of this contract are very generous.[17] For SNEPCO, the government reduced the petroleum profit tax (PPT) to a flat 50 percent for oil concessions in deep-water areas, and also increased investment tax credit from 20 percent to 50 percent. Further, after royalty has been paid, SNEPCO can keep all oil production proceeds to itself in order to recoup its expenses before paying the PPT. Petroleum economist Sarah Ahmad Khan has warned that SNEPCO and other oil companies that are benefiting from the new production-sharing contract arrangement are now poised to take over Nigeria's oil reserves completely:

> While the oil companies have continually claimed that the new production-sharing contract incentives are necessary for investment in exploration and production in deep-water acreage, it does seem that the government has been very generous to the oil companies in setting up the initial tranches of profit-sharing oil. For instance, an oil company that discovers only one large field of about 40,000 barrels per day would have access, once tax is paid and cost recovered, to 80 percent of the fields of production over a period of ten years. Given the financial straits that will continue to constrain the country in the medium term, these production-sharing terms can be seen as a signing away of Nigeria's reserves and future production for a significant period.[18]

In dollar terms, this means even more profit for Shell. But even this does not tell the whole story. Nigerian journalists who have investigated the company's activities in the country over the years have come to the conclusion that Shell's real profit is probably far greater than the figures it presents to the federal government's auditors and accountants for tax assessment at the end of the financial year and that the bulk of its earnings, over and beyond the one dollar per barrel it shares with Elf and Agip, comes from an elaborate mechanism the company has put in place to sideline the provisions of the MOU, leaving the government with the short end of the stick. While Shell has vigorously denied these allegations, a panel set up by the new Nigerian government in 2000 has been uncovering a veritable can of worms. The respected Nigerian daily *ThisDay* reported on indications that Shell was negotiating with the Nigerian National Petroleum Company over a substantial refund.

This is how such sidestepping works: A 1972 enactment, "Regulations in Employment of Nigerians," made it compulsory for all oil companies operating in the country and wanting to dispense with the services of Nigerians in their employ and replace them with expatriate staff, to first seek approval from the Ministry of Petroleum Resources.[19] The ministry, before granting the permission, would verify that there were no qualified Nigerian personnel capable of doing the job and that the company was not exceeding its expatriate staff quota. Thus, while this law was still in force, it was virtually impossible for the oil companies, including Shell, to do away with their Nigerian staff and replace them with expatriates without real cause. A memo from the Office of the Petroleum Minister in January 1991 changed all that, however.[20] The memo empowered the oil companies to dispense with the services of Nigerians in their employ without clearing this with the Ministry of Petroleum Resources, and also gave them permission to hire expatriate staff without referring to National Petroleum Investment and Management Services (NAPIMS), the NNPC subsidiary that monitors and regulates the activities of joint venture partners.

Ademola Adedoyin, a Nigerian journalist who has been monitoring the activities of the oil companies, says that Shell and the other joint venture partners have been flouting the expatriate quota since the January 1991 memo. For the federal government, he said, "this is double jeopardy, as it has ended up picking up heavy bills as salaries and allowances of the expatriates."[21] In January 2000, a Shell employee, Emeka Nwawka, an engineer, lodged a complaint before the House of Representatives Committee on Public Petitions, "alleging that Shell was terminating his appointment in order to replace him with expatriates. Nwawka, who also alleged expatriate quota violations against the oil giant, had also gone to court and obtained an injunction restraining Shell from terminating his appointment until the House concludes its investigations and passes a resolution on the matter." Shell responded by filing a suit at the Federal High Court on February 14, 2000, asking that the House Committee be compelled to suspend all action on the matter until the court proceedings were over. The case is pending in the courts. It is instructive that Shell's concern in heading for the law courts was not to challenge Mr. Nwawka's allegations on expatriate quota violation, but indeed to stop the House Committee's investigations into this and other alleged violations.[22] Adedoyin further explained that the expatriate staff earn their salaries in hard currency paid into their accounts abroad, while they are also paid a living allowance as long as they are in Nigeria. This allowance is almost equal to the total remuneration of their Nigerian counterparts. "The implication of this is that government, as the senior partner in the joint venture arrangement, picks up about 60 percent of this bill, or 55 percent in the case of Shell."

By the revised 1986 MOU, Shell is permitted to deduct $2.50 from every barrel sold, as costs. In reality, however, the bulk of this cost is borne by the federal government because it pays 55 percent of the salaries and emoluments of Shell's expatriate personnel, who ostensibly produce the oil, and this further boosts the company's profits.[23] Discussions are presently under way between the government and the joint venture partners to raise the notional technical cost per barrel to $2.90. Indeed, what Shell and the other oil companies claim to be actual technical costs is higher than $2.90. It ranges between $4.50 and $6, and varies from company to company.[24] Shell has argued that easily accessible oil reservoirs are becoming fewer, and the need to employ more sophisticated technology to explore for oil in more difficult terrain invariably means that production costs would rise.[25] But Ademola Adedoyin has written that exploration costs in the Gulf of Mexico and the North Sea are nose-diving as a result of improved technology and efficiency, while the average unit production cost for Nigerian oil fields has been rising progressively. This, according to him, means that Shell as the operating company has the government in a viselike grip. The higher the production cost that Shell claims, the lower the Petroleum Profit Tax (PPT) it pays to the government. Also, the higher the production cost, the higher the amount the government pays as cash call.

Energy economist Jedrzeg George Frynas has also argued that the control over operating costs is probably the key to the understanding of high profits in the country's oil industry. He quotes a former chief executive of the NNPC, who confessed that "proper cost monitoring of their operations has eluded us, and one could conclude that what actually keeps these companies in operation is not the theoretical margin, but the returns which they build into their costs."[26] Noting that the operational budget in Nigeria is decided by Shell, Frynas argues that the company has a financial incentive to inflate costs.

The NNPC, as we have pointed out, does not have the technical expertise to verify the authenticity of these production cost claims. While it must be admitted that corruption in government circles ensures that the bulk of the oil revenue simply disappears into private bank accounts of senior members of the government and their hirelings, the ever-rising production claims put forward by Shell and the other joint venture partners has put the NNPC in a very difficult position—it now finds it increasingly difficult to meet its financial obligations as the senior partner in the venture. In 1994, Shell claimed that the NNPC owed it a total of $380 million as exploration and production costs. Chevron said it was owed $200 million, Mobil $180 million, and Elf and Agip $10 million each. By early 1995 the total amount claimed by the oil companies had risen to $1.1 billion. Significantly, government officials argued that they were prepared to acknowledge only a debt of $400 million. Following a series of negotiations, NNPC agreed to a debt settlement of $625 million and was asked to pay this to the oil companies before the end of 1996.[27]

In September 1996, however, Nigeria's Minister of Petroleum Resources, Daniel Etete (who has since been replaced), called a press conference and declared that henceforth his ministry would monitor properly and "very thoroughly" all cash call claims by the joint venture partners. Etete said there would be established, in the office of the Minister of Petroleum Resources, "a Monitoring Unit, which shall veto all invoices and claims to be debited to the cash call escrow accounts." Etete also said that all the oil companies would be asked to reapply for oil licenses under a new guideline obtainable from the Department of Petroleum Resources (DPR).[28] The federal government was alarmed that the production costs of Shell and the other oil companies were ever on the increase. The monitoring unit does not have the necessary technical expertise to verify Shell's claims, and so every year the NNPC is saddled with huge production and cash call bills. And NNPC's loss is Shell's gain—running into millions of dollars every year.

In an interview with *Newswatch* magazine in September 1996, Etete implicitly suggested that Shell and the other oil companies operating in the country were shortchanging the government and people of Nigeria. Said Etete in the interview, "I have taken stock of our oil industry. I have looked at the nation's share from the oil industry over the years vis-à-vis the major operators in the industry, that is, the oil companies. From the evidence before me, I feel sad that our country has not received a fair share from its own resource." The minister explained that although the country had a 60:40 percent equity agreement with the oil companies, government officials are not informed of the other assets accruing to the country apart from the oil. Said the minister, "There should be other assets—fixed assets, and equipment and so on and so forth. But in this case of our venture partnership with the operators, we are not aware of all these things. The oil companies are the sole signatories to the accounts. Is that a fair deal?" Etete also leveled other accusations against Shell and the other oil companies. He said they regularly overspent their budget, as agreed under the terms of the MOU, without the authorization of the government, and then turned to the latter to pay 60 percent of the excess. He complained that the federal government picked up virtually all the expenses of the oil companies, including the coffee and tea the workers drink, and also the expenses for hiring helicopters, which could cost as much as $2,500 an hour. Etete asked rhetorically, "How much do they [Shell and the other oil companies] invest and how much do they take home? Let us be honest with ourselves. How much comes to government, how much goes to the oil-producing communities as against the huge amount of money they take away?"[29]

As the 1996 Greenpeace report rightly concluded, "It is not easy to calculate how much the private oil companies earn on each barrel of oil."[30] Going by the revised MOU, Greenpeace estimates that Shell earns an average $500 million profit in Nigeria every year, of which $300 million is reinvested in the country and the rest sent to the Group's headquarters in London and The Hague as dividend for shareholders. The Greenpeace figures merely scratch the bottom of Shell's profit barrel, as the country's petroleum resources minister has shown. The NNPC's inspectorate division suffers from a chronic dearth of technical staff and so cannot verify Shell's claim that it spends well over the $1.50 minimum on every barrel. And this, significantly, has been a major source of disagreement between Shell and the Nigerian government as the company demands more and more money from the government in the form of cash calls and "rising production costs."

There is also the issue of exactly how much Shell spends on its so-called community assistance projects in the oil-producing areas, expenditures for which it claims tax relief. These expenditures have not been independently verified to ensure that the figures put out by Shell officials for tax purposes is an accurate reflection of what has actually been spent to help develop the communities. Then there is the pittance the company pays to the communities for "fair and adequate compensation," saving millions of dollars every year in the process. When the foregoing is added to Greenpeace's $500 million a year, the true picture of Shell's earnings in Nigeria begins to emerge—and $1 billion per year is not too wide of the mark. Little wonder, then, that Mark Moody-Stuart, Shell's chairman, would remark to a *Financial Times* journalist in February 1999, "Nigeria is fundamentally a low-cost oil producer. It is strategic to the future of the group."[31]

By Shell's own admission, the company has, to date, extracted 634 million barrels of oil from its ninety-six oil wells in Ogoni alone. Shell also claims that before it withdrew from the area in January 1993, Ogoni accounted for 1.5 percent of its Nigeria production, down from a peak of

5 percent in 1973.[32] And yet from this relatively small community, with a total land area of 400 square miles, it's estimated that Shell has extracted over $30 billion worth of oil.[33] Shell claims that its community development programs in Ogoni and the other Niger Delta communities date back to 1958 and that it has increased its efforts in recent years and is now spending an average of $20 million every year (down from the $25 million it claimed in 1994). Local NGOs in the area dispute these figures, arguing that between 1970 and 1988 the company spent only an estimated $200,000, or just 0.000007 percent of the value of oil it extracted from the Niger Delta, in its community projects in the region.[34]

Clearly, then, the relationship between Shell and the communities is not anywhere near the nice and cozy "partners in progress" picture so assiduously promoted by the company's image makers.

Sokebolou and Other Tales of Woe

Shell has found the various legislation enacted by the federal government, particularly Petroleum Decree No. 51 of 1969 and the 1978 Land Use Act, a useful shield with which to fend off criticisms of neglect leveled against it by the communities and other concerned interest groups. The company's stock response is that it pays royalties and rents to the federal government amounting to more than 90 percent of the net oil revenues, while it also pays compensation to the communities for the surface rights of all land acquired in the course of its exploration and production activities, and for damage, including oil spills. Shell says its compensation rates are fair and adequate, over and above the statutory minimum stipulated by the federal government, and it has reason to believe that all the parties concerned, including the communities themselves, are happy and satisfied.[35]

The communities have countered that these cruel and morally reprehensible decrees that took over their land and resources by force were enacted by military dictators who did not consult them, and that Shell supports and indeed profits from their collective misery. They also point out that even in such relatively simple matters as compensation for land and other private property taken over or destroyed by Shell in the course of its operations, they are subjected to humiliating treatment by company officials. And they complain that more often than not, Shell has to be compelled by the courts before it pays these compensations, and then after long and frustrating delays, when inflation would have eaten deep into the value of the money. In September 1997 the people of Ekeremor Zion, a small impoverished community in the western Delta, finally won a court case that they had instituted against Shell along with several other communities in 1988. Instead of paying the community compensation for the loss they had suffered, the company decided to appeal to a higher court. Shortly after Shell took this decision, Nigerian soldiers, armed to the teeth, moved into Ekeremor Zion and razed the village to the ground. The community is still mourning its dead.[36]

"We Are Suffering, We Are Dying"

Shell rarely pays compensation until it is compelled to do so. And even so, Shell has no policy to pay fair compensation. The celebrated case between Farah and Shell Petroleum Development Company Ltd. in 1995 illustrates this point. After an oil spill incident in K-Dere, Gokana-Tai Eleme Local Government of Rivers State, in 1970, in which Shell was implicated, it took twenty-five years for the plaintiffs who took Shell to court to get adequate compensation.[37] Elsewhere in the Niger Delta the general complaint is one of broken promises, development assistance programs that are abandoned halfway, and poor-quality facilities that break down and simply rust away as soon as they are installed.

Yenogoa, capital of the newly created Bayelsa State, produces some 40 percent of the nation's oil and yet has no pipe-borne water, no hospitals, and motorable roads, not even electricity mains linked to the national grid.[38] Since it opened its Utapete flow station in Iko in 1974, Shell has extracted an estimated $1.5 billion worth of oil from the village. The community, on the other hand, estimates it has sustained losses up to $300 million as a result of Shell's activities in the area.[39] Anietie Usen, a Nigerian journalist, visited the community in December 1985 and reported: "Iko, that wretched oil-rich village. That shady village of oil and palm trees. Iko, a reporter's nightmare, the oil man's goldmine."[40]

On October 31, 1995, shortly after the Nigerian military junta sentenced Ken Saro-Wiwa to death, Shell rushed out a press statement claiming it had spent what amounted to about $720,000 in its Shell East Community Assistance Projects, including Ogoni, between 1986 and 1993.[41] A team of NGOs and community groups comprising Ogoni Community Association U.K., Civil Liberties Organization, Niger Delta Watch, Centre for Nigerian and International Environmental Law, and Environmental Rights Action (ERA), however, visited Ogoni with a view to authenticating these claims and discovered that they were to a very large extent exaggerated.[42] Shell, in response, disagrees that the total value of the oil it extracted from its Ogoni wells between 1958 and 1993 when it pulled out from the area is $30 billion. The company has valued the oil taken from the area before it pulled out in 1993 at $5.2 billion, claiming that 79 percent of this went to the Nigerian government in taxes, royalties, and equity take. While Ogoni leaders insist on $30 billion as nearer the mark, it is important to note that no independent financial audit of Shell's Ogoni operations from 1958 to 1993 has been undertaken to confirm the company's $5.2 billion figure.

All over the Niger Delta, the complaints and grievances are shockingly similar—villages and whole communities pining away in poverty and neglect under the intimidating shadow of a bloated Shell. The struggle of the Nembe people against a monstrous multinational that takes away their oil and gives them little in return began in September 1990. Obasi Ogbonnaya, the journalist who visited Nembe and Okoroba in March 1994, was surprised to discover that virtually every youth in the area had the facts and figures of Shell's operations in Nembe Kingdom and environs. Explained Ogbonnaya, "The people use these figures to justify their militant opposition to decades of neglect, systematic impoverishment, pollution of land, water, and air, destruction of fishing nets and of communities."[43]

In Okoroba, a few miles from Nembe, where Shell also operates an oil well, the inhabitants are still reeling with pain after the company's dredgers went through the village, ravaging economic crops and farmlands, practically destroying the human ecosystem. Many families who lost their farms starved the following year. Shell paid a paltry two million naira ($20,000), in total, to the over six thousand people who lost everything to its dredgers, from food to irreplaceable cultural artifacts, including the graves of loved ones.

Ronnie Siakor, an environmentalist and community worker who lives and works in the Niger Delta, was contracted by Living Earth (a London-based environmental NGO) to visit some selected communities in Shell's Western Division operations in April 1996. The trip was sponsored by Shell

International Petroleum Company (SIPC) in London, ostensibly to provide Living Earth with background material for pilot educational development projects the NGO wants to set up in the communities of the Niger Delta in collaboration with the multinational. Shell paid Living Earth $96,000 to conduct the preliminary survey.

The Living Earth survey was brief, lasting only a few days. Still, Ronnie Siakor saw enough of Shell's "development" projects in Sokebolou, Ogborodu, Agidiama, and several other communities in the western Delta to make him livid with anger. Said Siakor in his report: "Shell tries to establish self-help development projects in the communities, but according to our investigations it is really Shell-help."[44]

"The Black Hole of Corruption"

On Sunday, December 17, 1995, the *Times* of London reported that Shell was planning a purge of executives in Nigeria "following the discovery of a 'black hole of corruption' involving the payment of millions in bribes and kickbacks to tribal chiefs, community leaders and the military in the troubled Ogoni region."[45] It is interesting that it took the brutal murder of an eminent writer and political activist and his eight compatriots for the *Sunday Times* Insight Team to beam its searchlights on the Niger Delta and tell the international community what Shell had been doing in the area for the past four decades.

Of course, for the inhabitants of the over eight hundred communities where Shell has its operations, the company has long been associated with charges of corporate corruption and double-speak. Steve Lawson-Jack, former head of Shell's public relations and governmental affairs in the Eastern Division, had over the years perfected the art of speaking from both sides of his mouth, promising the communities development assistance that never materialized. The public relations "expert," however, became a public relations liability for his employees when Saturday Kpakol, a prominent Ogoni leader, alleged that Lawson-Jack had offered to split a 500,000 naira contract ($5,000) with him in return for working to subvert MOSOP.[46]

An audit team investigating Shell's Nigeria operations also "discovered" that Lawson-Jack was involved in a $100,000 compensation claim against the company for a nonexistent oil spill. The managing director of Shell in

Nigeria at the time, Brian Anderson, conveniently went public and distanced his company from its image maker and government point man, declaring that he was considering Steve Lawson-Jack's future and that twenty other Shell employees could be dismissed. Said a "contrite" Anderson, "It's like a black hole of corruption, acting like a gravity that is pulling us down all the time."[47] (Anderson later denied the statement credited to him by the *Sunday Times* concerning Lawson-Jack.)

One area where the company has ample room to cut corners is compensation claims, where it has found the country's inchoate oil mineral laws a most useful ally. While government legislation is clear and unequivocal about ownership of land, oil, and other minerals, vesting them in the federal government of Nigeria through Petroleum Decree No. 51 of 1969 and the 1978 Land Use Act, there are no clearly defined laws guiding compensation claims in the oil industry. Going by the foregoing decrees, Shell is given an oil mining lease by the federal government, and is not legally obliged to pay any compensation to the local communities on whose land it explores and produces oil. However, the company is required to pay fair and adequate compensation to the communities for the surface rights of all land acquired. It is also required to pay compensation for damage that results from spillage and other related incidents.

It is in the interpretation of the phrase "fair and adequate compensation" that Shell uses its awesome clout to exploit the communities. The process of lawmaking in Nigeria is notoriously slow, bogged down by bureaucratic red tape. Thus, the statutory minimum rates of compensation set by the General Babangida regime in 1987 are ridiculously low. According to government regulations, Shell is expected to pay about twenty-five cents for every mango tree it uproots in the course of its operations, though an average mango tree in Nigeria today produces an estimated $800 worth of fruit every year and has a life span of some fifty years. Its stem and branches are also a valuable source of timber, yielding an average $1,600 each year. Even Shell officials, obdurate and uncaring as they are, recognize that twenty-five cents for a mango tree is not a bargain, but plain robbery, and claim to have unilaterally increased the government-approved rates in June 1992. The Federal Military Government, faced with increasing unrest in the oil-producing communities following Ken Saro-Wiwa's murder, caved in to public pressure and upwardly reviewed the compensation rates in late 1997.

But what does this so-called increment amount to? In any case, the compensations, when they are paid—and this is very rare—usually end up in the pockets of corrupt community leaders as payment for services rendered to Shell, holding down their people so the company can extract all the oil it desires without being challenged. Said a European Shell executive, commenting on his company's activities in Nigeria, "I would go so far as to say that we spent more money on bribes and corruption than on community development projects."[48]

Shell claims in its briefing notes that "allegations that we take advantage of 'illiterate landowners' in negotiations are insulting to the communities," and that in its opinion the people with whom it negotiates were well-educated and aware of their rights. Obviously, Shell does not intend this statement to be taken seriously, because elsewhere in the same briefing notes it acknowledges that communities in the Delta area "still lack basic facilities such as electricity, running water, roads, sewage treatment facilities and have limited opportunities for education and employment."[49] As Andrew Rowell, the British environmentalist who has written extensively on the communities of the Niger Delta and their struggle for social and ecological justice, has written, "Three-quarters of Ogoni cannot read or write and cannot understand the compensation forms."[50]

The present compensation process is very complicated, and involves negotiation between affected communities and the oil companies as the polluter or purchaser through their various lawyers. If both parties fail to agree on a schedule of compensation, the Department of Petroleum Resources is expected to intervene and mediate. If this also fails, as is often the case, the two parties end up in court. And it is here that Shell comes into its own. The overwhelming majority of community leaders are poorly educated and often commit themselves to agreements whose legal implications they do not fully understand until they get to court. Moreover, the communities themselves are usually poor and in no position to afford the expensive legal fees. Frequently they're forced to settle out of court, accepting whatever handout Shell officials choose to give them.

Ideally, rates of compensation payable for crops and trees that are damaged or affected by oil exploration activities are determined using the approved government rates and such other considerations as court judgments, crop yield researches, and the reports of independent valuers commissioned by Shell itself. Procedures for compensation are, however, notoriously chaotic. Indeed, a policy document written by a technical committee and designed to streamline assessment of damages due to oil pollution has yet to receive the attention of the federal government. It is also significant that as late as 1995, Nigerian courts were still delivering judgments, as in the *Shell* v. *Farah* case, that "serves as a beacon of light to oil mineral producing areas of Nigeria and a guide to oil companies in compensation claims."[51] The communities are too poor to employ the services of their own valuers independent of Shell. Neither do they have access to the latest reports of crop yield researches—that is, where they exist at all. Given all these constraints and inadequacies, Shell, supported by government officials, simply treats the compensation claims as it pleases.

What If Shell Left?

What would the communities of the Niger Delta lose if Shell were to pack up its oil rigs today and leave? Nigeria earns over 95 percent of its export income from oil, receiving an estimated $20 million daily from sales. However, as Nick Ashton-Jones has pointed out, a simple cost-benefit analysis would indicate that rather than contribute to the social and economic well-being of the communities, the oil industry as it presently operates in the Niger Delta is a net disbenefit.[52] Among the direct costs are environmental degradation resulting from pollution (offshore and onshore spillage and gas flaring), the destruction of the natural hydrology of the area through the construction of poorly designed canals to facilitate transport of men and materials, and the slow but relentless havoc wreaked on the ecosystem by such activities as seismic testing, road construction, well drilling, and pipeline laying. There is also the concentration of lethal gases in the atmosphere, the introduction of harmful acids into otherwise fertile soils, and the pollution of groundwater by effluents.

Andrew Rowell has also drawn attention to the fact that the Niger Delta is sinking gradually due to oil operations. Citing a World Bank report, he says an estimated 80 percent of the population will have to move as a result of sea-level rise, and that the resulting damage to property will be in the neighborhood of $9 billion. Rowell paints a frightening but very real scenario in which the sea is rising while the Niger Delta itself is subsiding, and quotes a report by community leaders in the area, estimating that a twenty-five-mile-wide strip and its people could be washed into the sea in the next twenty years.[53]

With the advent of the oil industry in the late 1950s, the Niger Delta witnessed an influx of people from other parts of the country—all looking for jobs in the oil companies.[54] Thus, from a modest 76,000 people in 1952, the population of Port Harcourt, the chief city of the area, tripled during the oil-boom 1970s. Today it is a sprawling conurbation of over a million inhabitants. But the "oil" jobs are few and far between. According to the World Bank, Shell and the other oil companies invest some $30 million in oil-related activities in the Niger Delta every year, but their initiatives to improve the quality of life in the oil-producing communities have been "minimal."[55] Meanwhile the rapidly expanding population in Port Harcourt and the other towns and villages have to eat, and they turn to the land, a good part of which is flooded during the rainy season. Overfarming, coupled with Shell's devastation of the environment, soon left the available cultivable land gasping and devoid of nutrients, and the creeks and rivers stripped of their fish population. Hunger leads to anger, and the crushing poverty and marginalization of the communities, in contrast to the oil resources that are rightly theirs, provide the trigger. A war of all against all ensues: youths against elders, whom they accuse of selling them out to Shell; community against community, in competition for scarce Shell contract work; and communities against Shell and the federal government, who deny that their actions have driven the people of Nigeria into a dark, impossible corner.

In their defense, Shell officials argue that it is not fair to blame only their company for the social and environmental problems of the Niger Delta, pointing out that "Mobil, the second largest producer, produces from offshore fields, and in many ways is 'out of sight and out of mind.'"[56] They also say that if Shell is "punished" by the international community and forced to withdraw from Nigeria, the people of the Niger Delta would be worse for it, as no other oil company operating in the country presently (including others that the government might invite from outside to take over Shell's concessions) can match Shell's health, safety, and environment standards.

Shell is right—but only to the extent that the other oil companies are equally culpable in the destruction of the Niger Delta human ecosystem.

Shell's discovery of oil in Oloibiri in 1956 quickly drew other multinational companies to the Niger Delta. The federal government granted two

Oil Exploration Licenses (OEL) for the country's continental shelf to Mobil, Texaco, and Gulf (now Chevron), respectively, in 1961. Shell-BP, which already held an Oil Mining Lease (OML) on 15,000 square miles of Nigeria's land area, was also given four OELs to prospect for oil offshore.[57] Chevron began to produce oil from its offshore fields in 1965, Elf (former Safrap) in 1966, Mobil in 1969, and Texaco and Agip a year later.[58]

It could be argued that these oil companies that came later simply borrowed a leaf from Shell's book, the biggest oil company in the area, and adopted a lackadaisical attitude toward the environment and the general welfare of the local communities. The World Bank has reported that they use even worse methods than Shell in the treatment of water generated in the course of oil production.[59] They also spill oil and flare gas indiscriminately. One year after a storage tank in Shell's Forcados terminal ruptured in 1979, spewing 570,000 barrels of oil into the estuary and the adjoining creeks, a blowout in Texaco's Funiwa well discharged 400,000 barrels of oil into the coastal waters and laid waste to 840 acres of mangrove. It is reported that 180 people died in one of the communities severely affected by the resultant pollution. Agip's Ogada-Brass pipeline is also notorious for spills—a regular occurrence since 1988, when 10,000 barrels of oil escaped from the pipeline and polluted the surrounding vegetation.[60]

Mobil is also a culprit in this regard, coming closely after Shell. An estimated 20,000 barrels of oil were spilled into coastal waters in 1995 when the company's production platform located twenty-two miles offshore (near Ibeano) exploded, also claiming ten lives.[61] On January 12, 1998, the twenty-seven-year-old pipeline in Mobil's Idoho platform leading to the Qua Iboe terminal ruptured, spilling some 40,000 barrels of light crude into the sea. The spill quickly spread, damaging fishing nets, polluting farmlands and water sources, and triggering water-borne diseases in at least twenty-two villages with an estimated population of one million.[62] Mobil, along with two other oil companies, has also been taken to court by some communities because of a substandard Environmental Impact Assessment. Reenacting Shell's Okoroba disaster, Chevron dug a canal in Awoye village in Ondo State, opening the area up to coastal erosion, causing salt- and seawater areas to mix, and wiping out traditional fishing grounds and sources of drinking water. Bruce Powell, an authority on the Niger Delta human ecosystem, described the resultant damage as "one of the most extreme cases of habitat destruction" in the Niger Delta.[63]

Shell does not have a monopoly on inviting armed soldiers and police to restore "peace" in its areas of operations. As Human Rights Watch has reported, Chevron called in the police after protesting youths at Opuekebo in Delta State blockaded its facilities in the area with sixteen boats strung together in May 1994. The police deployed a self-propelled barge, which rammed the blockade and sank the boats.[64] Antiriot police had stormed the Obagi community, which accounts for 70 percent of Elf's total oil production in the country, the previous February, after the people complained to company officials that they were not getting any worthwhile benefits from the oil taken from their land. Their spokesman, Professor J. G. Chinwah, was falsely accused of murder and detained.[65] However, for all the lapses and sundry misdemeanors of these other companies, it is Shell, as the operating officer of the largest oil-producing venture in Nigeria, that sets the standards for the others to follow. Shell officials are therefore disingenuous when they raise the possibility of being replaced by other companies with even worse standards. The question is not who is the greatest polluter among the oil companies presently operating in Nigeria. It is about getting Shell officials to treat the people of the Niger Delta and their environment with proper consideration.

Would life come to an end for these communities if Shell were to suddenly pack up and go, as it has suggested in its publicity campaigns? Let us assume, for the sake of argument, that one immediate repercussion of the company's departure would be the temporary loss of the oil asset and the resultant disruptions in community life this would cause. Competition for scarce cultivable land and other economic resources would increase as the immigrant population who had been lured to the area by the prospects of jobs in the oil industry turned to other activities in order to survive. The facilities abandoned by Shell would also constitute a health hazard if they were not properly disposed of. Life, however, would certainly not come to an end if the oil wells were to be shut down permanently. Attention would gradually return to the long-neglected Renewable Natural Resources sector (agriculture, forests, and fisheries)—the basis of real sustainable wealth. Improvements in soil conditions and farming methods would be assiduously pursued, and agricultural productivity would increase to sustain the expanding population in the Niger Delta.[66]

Let us consider, on the other hand, the benefits to the communities if Shell were to operate in a socially and environmentally responsible way and respected the property rights of the communities. The federal government would let the communities have a fair share of the oil wealth, and there would be jobs for them in Shell and other ancillary industries, which in turn would generate social amenities that would enrich the lives of the inhabitants.

But all these are theoretical scenarios, and we are living in a real world in which Shell, even now, is ravaging the human ecosystem of the Niger Delta beyond repair. As Ashton-Jones has argued, a proper cost-benefit analysis might suggest a net benefit or a net disbenefit, but the real test would be to assess what situation the oil-producing communities would be in after Shell had extracted all the oil, damaged the environment, and departed. An economic asset would have disappeared without the communities getting a share of the wealth it generated—their land and rivers and creeks damaged beyond repair, leaving them with nothing to fall back on for sustenance, and with a local economy disoriented and reeling with shock after the oil industry that had powered it for so long had crumbled. Said Ashton-Jones, "The economic costs might be deemed acceptable if oil revenue was, on the whole, remaining in the oil belt to create a viable agricultural and industrial economy: it is not. The costs might be deemed acceptable if the oil revenue was being invested outside to produce an income in the future: it is not."[67]

Oppressed, repressed, and denied their property rights in turn, the oil-producing communities of the Niger Delta have become living carrion on which successive regimes in Nigeria and their foreign collaborators, like insatiable vultures, have feasted, and are still feasting, without letup. This particular carrion would have nothing to lose if its tormentors were suddenly to leave her alone. It has its life to gain.

SIX

Ambush in the Night

The sound of gunfire in the streets brought Maria Nwiku racing from her home. As she emerged, a bullet to her leg sent her sprawling. Two of her children, running with her, were shot dead. As Nwiku watched, attackers forced their way into her house and murdered her elderly husband and their third child.

Newsweek, reporting the sacking of Kaa,
an Ogoni village, by Nigerian troops in September 1993[1]

From the dark days of slavery to the present, the Niger Delta has been ruled by violence, and men of violence have sought to rule her by force. The area's substantial natural and human resources have always proved an irresistible attraction for slave traders, commodity merchants, colonialists, and plain fortune hunters who subjugate the inhabitants through treachery and force of arms and plunder their resources. With the discovery of oil in the area in 1956 by Shell, the oppression and exploitation of the peoples of the Niger Delta entered yet another, and even more insidious, phase.

The Movement for the Survival of the Ogoni People (MOSOP) emerged in August 1990 to put to an end this dark chapter in the Niger Delta story. In the words of the writer and activist Ken Saro-Wiwa, "The Ogoni took stock of their condition and found that in spite of the stupendous oil and gas wealth of their land, they were extremely poor, had no social amenities, that unemployment was running at over 70 percent, and that they were powerless, as an ethnic minority in a country of 100 million people, to do anything to alleviate their condition. Worse, their environment was completely devastated by three decades of reckless oil exploitation or ecological warfare by Shell."[2]

In October 1990 the chiefs and community leaders of the six Ogoni clans came together at Bori and presented the Ogoni Bill of Rights, a document they had collectively adopted two months previously, to the government and the people of Nigeria. The OBR demanded, among other things, the right of the Ogoni people to self-determination as a distinct people in the Nigerian Federation; adequate representation as a right in all Nigerian national institutions; the right to use a fair proportion of the economic resources in Ogoniland for its development; and the right to control their environment. The OBR also emphasized that the MOSOP was a nonviolent organization, and believed in the use of nonviolent means to pursue its goals.[3]

The Ogoni are a "mere" 500,000 in a Nigeria with a population of over a 100 million people, dispersed in over two hundred nations and ethnic groups. Thus the launching of MOSOP did not even register on the national canvas save for a brief mention in some of the local newspapers in Rivers State. However, things began to change when MOSOP leaders on November 3, 1992, acting on behalf of the Ogoni people, issued a thirty-day ultimatum to all the oil companies operating on their land—Shell, Chevron, and the NNPC—to pay back-rents and royalties and also compensation for land devastated by oil exploration and production activities, or leave.[4] The memorandum, addressed to Shell, demanded the following:

a. Six billion dollars as unpaid royalties
b. Immediate stoppage of environmental devastation of Ogoniland, with particular reference to gas flaring at Yorla, Korokoro, and Bomu
c. Burying of all high-pressure oil pipelines currently exposed in all of Ogoni
d. Payment of $4 billion as reparation for damages and compensation for environmental pollution suffered by the people and their environment
e. Dialogue between representatives of the community, Shell, and the federal government

The three companies, like the government two years previously, ignored the demand. But they had not counted upon Ken Saro-Wiwa's organizational genius. A consummate publicist who had honed his craft writing novels, newspaper articles, and best-selling soap operas for the government-owned television network, Saro-Wiwa, in collaboration with other MOSOP leaders, had quietly embarked on a mass mobilization of Ogoni men, women, and children shortly after the movement was launched. The simple but ingenious innovations of the movement included the One Naira Ogoni Survival Fund, whereby all Ogoni people, young and old, were asked to contribute a token sum as an indication of commitment to the cause; and the formation of such pan-Ogoni organizations as the National Youth Council of Ogoni People (NYCOP), the Federation of Ogoni Women's Associations (FOWA), the Conference of Ogoni Traditional Rulers (COTRA), the Council of Ogoni Churches (COC), the Ogoni Teachers Union (OTU), the National Union of Ogoni Students (NUOS), Ogoni Students Union (OSU), Ogoni Central Union (OCU), and the Council of Ogoni Professionals (COP), for which MOSOP served as an umbrella. These ensured that the movement had a truly democratic, grassroots base.[5]

Ken Saro-Wiwa had always believed in the power of learning and the pen as instruments to help bring about progress and social change. From the onset, he urged his fellow Ogoni to study and to educate their peers about what MOSOP was really about—a movement for social and ecological justice, informed by the finest traditions of African participatory democracy and powered by the philosophy of nonviolence. Said Saro-Wiwa, "MOSOP was intent on breaking new ground in the struggle for democracy and political, economic, social, and environmental rights in Africa. We believe that mass-based, disciplined organizations can successfully revitalize moribund societies, and that relying upon their ancient values, mores, and cultures, such societies can successfully reestablish themselves as self-reliant communities and at the same time successfully and peacefully challenge tyrannical governments."[6] Saro-Wiwa's ultimate goal was a restructured Nigeria, functioning as a proper federation of equal ethnic groups and nations, irrespective of size, with each group free to control its resources and environment and also exercise its political right to rule itself according to its particular inclination.

The immediate task, though, was the strengthening of MOSOP, and even more important, the urgent need to take its case to the Nigerian people and the international community and find allies among them. Saro-Wiwa found sympathetic ears particularly among Nigerian journalists working in the independent press, of which he was considered a member. His talent for publicity was given free rein, and in a matter of months MOSOP and the travails of the Ogoni people became a subject of debate all over the country, especially in the early months of 1992. In his capacity as MOSOP spokesman, Saro-Wiwa traveled to The Hague in July 1992, where he registered the movement with the Unrepresented Nations and Peoples Organization (UNPO), whose charter enjoins nonviolence on all its members. He also brought the suffering of his people to the attention of the United Nations Working Group on Indigenous Populations in Geneva, and made useful contacts with international environmental groups and business organizations such as The Body Shop International, based in London, whose founder and chief executive, Anita Roddick, had long been involved in campaigns such as MOSOP was pushing in Nigeria.[7]

On January 4, 1993, approximately 300,000 Ogoni men, women, and children took to the streets and staged a peaceful protest against Shell's ecological war and the government's continued denial of the Ogoni right to self-determination and a fair share of their natural resources. The demonstration was timed to coincide with the start of the United Nations Year of Indigenous Peoples. This protest, so brilliantly organized that not a single incident of violence marred the event, marked a turning point in MOSOP's campaign and told the military government and the Nigerian people in clear, unmistakable terms that a formidable new organization had entered the national political stage. On that day also, now marked as the Ogoni national day, the Ogoni people crossed the psychological barrier of fear and signaled to the military junta and its civilian allies that had been holding them down for over three decades that they were now prepared to take their destiny into their hands and liberate themselves from tyranny and oppression through nonviolent means.

General Ibrahim Babangida and the other members of the military junta were finally forced to take notice of MOSOP. A few days after the hugely successful January 4 demonstrations, the Inspector General of Police invited MOSOP leaders to a parley in Abuja.[8] Nothing came of the discussion, however. It was apparent the junta believed that a few harsh words would frighten the leadership into giving up their "dangerous" enterprise. Saro-Wiwa and the other MOSOP leaders were summoned to the headquarters of the dreaded State Security Service (SSS) in Abuja, where the riot act was read to them before they were sent away.

Unlike the military junta, however, Shell was monitoring MOSOP's activities closely, and its senior officials were sufficiently alarmed to initiate a strategy meeting between executives of Shell Nigeria and Shell International in Rotterdam and London in February 1993. Leaked minutes of the meeting indicated that Shell recognized that "the main thrust of the [Ogoni] activists now seems to be directed at achieving recognition of the problems of the oil-producing areas by using the media and pressure groups." The meeting also decided that officials of Shell Nigeria and Shell International should keep each other more closely informed to ensure that movements of key players, what they said and to whom, was more effectively monitored to avoid unpleasant surprises adversely affecting the reputation of the Shell Group.[9]

The first real confrontation between MOSOP and Shell came on April 30. Willbros, a U.S. pipeline contractor commissioned by Shell, was digging up newly planted farmland and laying pipelines in the village of Biara. The local farmers came out and challenged the Willbros workers, pointing out that they had not been paid any compensation for their land nor had a proper environmental impact assessment been conducted for the project, as stipulated by Nigerian law. A contingent of the Nigerian Army accompanied the Willbros workers. These soldiers subsequently shot at and dispersed the protesters. A young man, Agbarator Friday Otu, was killed. Eleven others received gunshot wounds.[10]

Following this incident, there were spontaneous peaceful demonstrations throughout Ogoniland to protest these destructive acts by Shell, Willbros, and soldiers of the Nigerian Army. Calm was restored when MOSOP's Steering Committee dispatched Ken Saro-Wiwa and two others to speak to the people. Shell subsequently claimed that it had ceased operations and pulled out of Ogoni because of public hostility to its activities.[11] In fact, not a single Shell worker was the recipient of "hostile acts" by MOSOP members, whose opposition was intended to empower their fellow Ogoni to stand up for their rights.

News that the oil giant had been "forced" out of one of its oil fields in the Niger Delta sent shock waves through the country's security apparatus. There was an immediate national alert, and references to "another Biafra" were routinely made in the security reports that streamed to Abuja from Port Harcourt, the Rivers State capital. On May 7, 1993, one week after the Biara shootings, MOSOP leaders were invited to Abuja to a meeting with the top echelon of the military junta's security establishment. The Ogoni were represented by Ken Saro-Wiwa, Dr. G. B. Leton, A. T. Badey, and Chief E. N. Kobani. (Badey and Kobani, along with two other chiefs, were later to be killed by an angry mob that accused them of collaborating with Shell and the government to subvert the MOSOP cause.) The junta was represented by Major General Aliyu Mohammed Gusau, National Security Adviser; Brigadier General Ali Akilu, Director of National Intelligence; and Alhaji Aliyu Mohammed, Secretary to the federal government.[12] The Ogoni leaders were asked to prepare a paper detailing their demands, a list of unemployed Ogoni youth, and a summary of the relationship of oil-producing communities in other parts of the world with their various governments and the oil companies. After the meeting, the four Ogoni leaders departed and nothing more was heard from the junta. To all intents and purposes, the Ogoni demands had been ignored.

Seeds of Discord

Shell was determined to return to Ogoni. The five oil fields were producing an estimated 30,000 barrels a day before the company announced in 1993 that it had pulled out of the area. Compared to the one million barrels Shell extracts from the other oil fields, this was a trifle. Shell officials, however, were anxious to see that the Ogoni "virus" did not spread to the other oil-producing communities in the Niger Delta, and so were determined to suppress MOSOP and use this as an example to other communities who might be tempted to tread a similar path in the future. Shell set about doing this with great cunning.

Shortly after the four Ogoni leaders returned from their trip to Abuja, Ken Saro-Wiwa embarked on yet another European tour to drum up support for the Ogoni cause. While he was away, Dr. Garrick Leton, MOSOP's president, and the late Chief Edward Kobani, the vice president, reportedly convened several public meetings in mid-May and attempted to convince the Ogoni people to allow Willbros, the Shell contractor, to resume laying pipeline.[13] When the people of a Gokana village, through whose land the pipelines were due to pass, sought reasons for this sudden about-face, they did not receive a satisfactory explanation. They subsequently refused to let Willbros onto their land. Saro-Wiwa returned from his trip on June 1, in time for a crucial meeting of the Steering Committee, where a motion to boycott the presidential elections scheduled for June 12 was to be debated. By now, however, it was clear that the cancer of discord had been introduced into MOSOP's body system by agents provocateurs.

Given the philosophical underpinning of MOSOP as reflected in the Ogoni Bill of Rights, the June 12 motion ought to have been a simple matter. MOSOP officials had been advised to shun party politics and the two political parties—the Social Democratic Party (SDP) and the National Republic Convention (NRC)—that General Babangida had created by military fiat. The constitution under which the presidential elections were being held also did not reflect the wishes and aspirations of the Ogoni people as contained in their bill of rights, and so the logical thing to do was to boycott the election and thus demonstrate to the government, and indeed the entire world, that MOSOP had no part in yet another election charade whose results would only perpetuate the regime of injustice and exploitation they were working peacefully to overthrow. Some progovernment community leaders, according to press reports at the time, saw things differently. They wanted Shell to resume operations in Ogoni, and MOSOP to participate in the presidential elections.[14]

These men, however, ran into a wall of opposition during the debate. Those who took the position that the Ogoni people should boycott the elections carried the day. Attempts were made to pressure Ken Saro-Wiwa into rescinding the decision of the Steering Committee, but he insisted that he would not be party to such an act. A few days later, Dr. Leton and Chief Kobani announced that they had decided to resign their positions as president and vice president of MOSOP. June 12 came, and the boycott was a resounding success, in spite of desperate attempts by some chiefs to lure the Ogoni into voting by sending out false information that Saro-Wiwa, who had traveled to Europe in the line with his publicity and diplomacy duties as MOSOP spokesman, had asked them to vote on the day.[15]

Afterward, Saro-Wiwa and the other MOSOP activists who believed, like him, that dialogue could only be initiated with Shell and the junta based on the demands of the Ogoni people in their bill of rights, became marked men. Guided by the resolution during the meeting of senior Shell executives in Rotterdam and London the previous February, Saro-Wiwa was followed everywhere by government and Shell security operatives and his activities closely monitored. (Irene Bloemink, of the Amsterdam-based environmental pressure group Milleu Defensie, has described how Shell officials monitored Saro-Wiwa's movements while he was on a visit to that city in February 1994, and even followed him into a meeting hall were he was to address Dutch environmental campaigners.)[16]

The Ogoni leader had previously been detained by the military junta, in April 1993, on frivolous charges. Following MOSOP's success in organizing a National Ogoni Vigil, a candlelight event to keep the struggle alive, attended by thousands, the military junta on May 2 enacted the Treason and Treasonable Offenses Decree of 1993, specifically equating secession with treason, punishable by death.[17] It became clear that the groundwork was being prepared for a major offensive against Saro-Wiwa and MOSOP. On June 21 he was arrested and detained, along with two other MOSOP activists, N. G. Dube and Kobari Nwile. Criminal charges were brought against them for belonging to MOSOP. While they were in detention in Owerri, matters came to a head in the movement as a faction led by Leton, who had earlier resigned his office as president, attempted to restructure MOSOP by suggesting that it cease to be an umbrella organization for NYCOP, COTRA, and the other subgroups. Leton and his group also leveled several allegations against Saro-Wiwa, among them that he sought to "hijack" the organization and also encouraged his supporters to employ "militant tactics." The rank-and-file members of MOSOP did not see any merit in Leton's case, however, and on July 6, MOSOP's Steering Committee elected Ken Saro-Wiwa president and spokesman of MOSOP in absentia. Ledum Mitee, a lawyer, was elected vice president.[18]

Shell's Cat Among the Pigeons

The Ogoni are a predominantly fishing and farming community who have always lived in peace and harmony with their neighbors—the Andoni, the Okrika, and the Ndoki. However, following the failure of the pro-government community leaders to "persuade" the majority of MOSOP activists to "see reason" with Shell, a plan involving security operatives in Rivers State working with directives from Abuja was hatched to cause mayhem in Ogoni under the guise of communal clashes. In July 1993, 132 Ogoni men, women, and children returning from a trip to the Cameroons were massacred on the Andoni River by uniformed men wielding automatic weapons. In August the Ogoni market village of Kaa on the Andoni border was attacked by a troop of men using grenades, mortar shells, and automatic weapons. Two hundred and forty-seven people were slaughtered, and the community primary and secondary school buildings were set upon and destroyed. Even as this grisly carnage was going on, the Ogoni villages of Tenama and Tera'ue, again on the Andoni border, were ransacked and several people killed.[19]

It was clear to Ken Saro-Wiwa and the other MOSOP activists who was behind these unprovoked attacks. The Ogoni had no dispute with their Andoni neighbors, and their territory was merely being used by forces intent on punishing the Ogoni for forcing Shell out of their land. It is inconceivable that the civilian governor of Rivers State at the time, Rufus Ada George, a former Shell executive, and the members of his Security Council did not know what was going on, but they acted as though the destruction of the Ogoni villages was just another communal clash. The governor set up the so-called Andoni-Ogoni Peace Committee, headed by Professor Claude Ake, to "mediate" between the two communities. There was, however, a change of government soon after, and the committee dissolved.

General Abacha, who was preparing to topple the Ernest Shonekan-led interim national government at this time, and needed all the support he could muster, invited Saro-Wiwa to Abuja. Abacha claimed he had been fed with false security reports on the Ogoni problem all along, and said he would move to remedy the situation and look into the grievances of the Ogoni. He apologized for the harsh treatment Saro-Wiwa had received at the hands of government security operatives and ordered that his passport, which had been impounded earlier in June, be returned to him. Back in Ogoni however, the plan to destroy the Ogoni had taken on a life of its own, nurtured by a deadly cabal consisting of some Shell officials and a handful of senior military officers and security operatives who were later to confess that they were in the oil company's pay.[20] Boats belonging to Shell were used by armed troops to attack several other Ogoni villages on the Andoni border in September 1993. A Bristow-owned helicopter that Shell usually charters for its oil production activities was also regularly sighted in Ogoni skies as these attacks were going on.[21] In the first two weeks of September over a thousand Ogoni were killed in the villages of

Eaken, Gwara, and Kenwigbara. An estimated twenty thousand more were rendered homeless.[22]

Professor Ake, who had been hoodwinked into heading the Andoni-Ogoni Peace Committee, now realized that other forces were at work and that there was more to these "disturbances" than met the eye. Thus, when the Rivers State Peace Conference Committee, composed of representatives of the Ogoni and Andoni, was hastily convened in October 1993 to broker an "accord" between the two communities under the auspices of a federal government agency, Ake dismissed it and called for an investigation to unravel the "mystery" of the sophisticated weaponry that was used by the Andoni, if in reality they actually had carried out the attacks on Ogoni villages. Said Professor Ake, "I don't think it was purely an ethnic clash, in fact there is really no reason why it should be an ethnic clash, and as far as we could determine there was nothing in dispute in the sense of territory, fishing rights, access rights, or discriminatory treatment, which are the normal causes of these communal clashes." Ake added that he suspected there were broader forces at work, which were interested in putting the Ogoni under pressure in order to derail the MOSOP agenda.[23]

A Human Rights Watch team was later to visit Nigeria, in the wake of the killings, and interview two soldiers who admitted that they had indeed been part of the army contingent that attacked Ogoni villages from Andoni territory. According to the soldiers, their units were instructed to assemble at a point in Andoni territory and then were informed that they were going on a mission to maintain peace between the two warring communities. On the way, however, the orders were suddenly changed and they were told to attack the Ogoni, who they were told were "causing all the trouble." Said one of the soldiers whom the Human Rights Watch report referred to as Corporal Number One: "I heard people shouting, crying. I fired off about one clip, but after the first shots I heard screaming from civilians, so I aimed my rifle upward and didn't hit anyone." The second soldier, who was part of a Nigerian contingent serving in the ECOMOG peacekeeping force in Liberia, also narrated how his unit was ordered home ostensibly to repel a Cameroonian attack. They were told to shoot on sight, only to later realize they were actually shooting at their fellow Nigerians—in this case unarmed Ogoni villagers.[24] Interestingly, all Ogoni policemen serving in the area were reassigned three weeks before Kaa and the other villages were set upon by the death squads. And when the attacks commenced, senior government officials in Port Harcourt pretended they did not know what was happening.

Attacks from Okrika and Ndoki territory in December 1993 and April 1994 respectively were similarly orchestrated. In the case of Ndoki, security operatives convinced members of the community to attack Ogoni villagers over a land dispute that had lingered for years but had never triggered any previous violent confrontation between the two groups. Then uniformed soldiers took over, ransacking eight Ogoni villages and killing everyone in sight. Two people from the Ogoni village of Barako told Human Rights Watch how they attended a town meeting in July 1994 and heard Lieutenant Colonel Paul Okuntimo, who was the commander of the Rivers State Internal Security Task Force at the time, denouncing MOSOP and boasting that he and his troops were responsible for the so-called communal clashes between the Ogoni and Andoni. One of the men recalled Okuntimo's words: "You [Barako villagers] are the worst type of people. You killed the Andonis. Then the Andoni let us know. So we came and chased you people. After the Andonis, you fought with the Okrikas and then with the Ndokis. So they invited us to chase you people. So we are the people who chased you from your houses and destroyed them."[25]

Paul Okuntimo's punitive raids from Andoni, Ndoki, and Okrika territory, vicious commandolike military expeditions in which thousands of Ogoni people were killed or rendered homeless, were a mere dress rehearsal for what was to follow. Meanwhile, the chief enemy of Okuntimo's alleged paymasters, Ken Saro-Wiwa, was still alive and causing "mischief."

Ken Saro-Wiwa: Chronicle of a Death Foretold

Ken Saro-Wiwa knew he had signed a pact with death when he sat down at his desk one early morning in 1990 and wrote the first draft of the Ogoni bill of rights. After he and other Ogoni community leaders launched MOSOP later that October, he began to talk incessantly about death. A man with a keen sense of history and imbued with a great intellect, he knew only too well what usually happens to the small guys who take on the juggernauts of this world. Said Saro-Wiwa, "When I decided to take the word to the streets, to mobilize the Ogoni people and empower them to protest the devastation of their environment by Shell, and their denigration and dehumanization by Nigerian military dictators, I had no doubt where it could end . . . death."[26] Like the great Nigerian musician, Fela Anikulapo-Kuti, Ken Saro-Wiwa carried death in his pocket.

Shell was still desperate to return to its oil fields in the Ogoni area. After the series of attacks on Ogoni villages by Lieutenant Colonel Paul Okuntimo, the oil company thought it was safe to return to Ogoni. And it went about this in an ingenious manner. First, Professor Isaiah Elaigwu, director of the National Council on Intergovernmental Relations, a government agency, was convinced to play a part in the supposed peace talks. Elaigwu arrived from Abuja as head of a "peace conference," ostensibly to settle the dispute between the Ogoni and the Andoni. But this was not just another peace conference. On October 4, less than two days after Eliagwu arrived, an "accord" was fashioned, supervised, conveniently, by Shell officials, who played a prominent role in what to many seemed an elaborate charade.[27] A document declaring that the discord had been amicably resolved and that all economic activity (meaning Shell's production activities in Ogoni) should resume was presented to Ken Saro-Wiwa to sign. He refused. When he later presented the document to the representatives of the Ogoni, they unanimously resolved they would have no part in Professor Elaigwu's so-called peace initiative.

Shell then unilaterally decided that there was an accord and ordered its workers back to work in Ogoni. Alarmed at the sight of Shell workers in their midst, the people of Korokoro protested nonviolently and asked them to stop work. Shell officials sent for soldiers. On October 23, Colonel Okuntimo, commander of the task force, arrived with a detachment of twenty-four military police in two buses, and then opened fire. A nineteen-year-old youth, Uabari Nnah, was killed and several others were wounded, including Papa Ndah, a seventy-year-old man. Shell officials witnessed the killing. Shell later claimed that it sent for soldiers because one of its fire trucks had been seized by the villagers, who also threatened its installations.

The truth, however, was that the fire truck incident occurred on November 3, many days after Shell officials called in Okuntimo and his soldiers. Meanwhile, Okuntimo, who had sustained a minor injury during the fracas, was taken to a Shell medical clinic, where he was treated at the company's expense. This sequence of events was later chronicled by the *London Observer*.[28] When Ken Saro-Wiwa learned of the Korokoro murder, he traveled to Port Harcourt and sought an audience with Brigadier General Ashei, Commander of the Second Amphibious Brigade, Okuntimo's immediate boss. Ashei sent for Okuntimo and, in Ashei's presence, Saro-Wiwa reportedly accused him of accepting blood money from Shell to kill Ogoni villagers.

Writing about this incident a few months later while he was in detention, Saro-Wiwa asserted that the Korokoro murder clearly indicated Shell's complicity in the state-sponsored violence that had been visited on the Ogoni since April 1993: "Shell is always there in the background even if it denies all participation. I believe, and not without reason, that the company's ready cash is always at play, goading officials to illegal, covert, and overt actions."[29] He did not know, however, that Okuntimo was plotting at this time to exact a most gruesome revenge for the Korokoro fiasco.

The Giokoo Murders

General Sani Abacha eventually struck on November 17 and assumed full powers as Head of State. Abacha dispatched one of his trusted military aides, Lieutenant Colonel Dauda Komo, to Rivers State as military administrator. One of the first things Komo did when he arrived at his new posting was arrest MOSOP's deputy president, Ledum Mitee, and Dr. Owens Wiwa, a member of the Steering Committee. Then, as Ogoni Day—January 4—was approaching, he placed Ken Saro-Wiwa and all the members of his family under house arrest. He did not lift the siege until January 5. Two weeks later Colonel Komo constituted the Rivers State Internal Security Task Force, composed of the army, navy, air force, Mobile Police, and State Security Service personnel, and appointed Major (later to be promoted Lieutenant Colonel) Paul Okuntimo, a former course mate of his at the Nigerian Defense Academy, its commander. This was the same Okuntimo who claimed to have acted in Shell's interest. Komo gave the task force one terse instruction: box in the Ogoni and subject them to the authority of the Rivers State Internal Security Task Force. Between them, General Abacha and Colonel Komo had finally worked out the solution to the Ogoni "problem," and it was left to Okuntimo to implement it.

The attacks on Ogoni villages from Andoni, Okrika, and Ndoki had left the land broken and bleeding. Fear and insecurity were pervasive, and villages tried to set up local vigilante groups to offer what little protection they could muster. There were also accusations and counteraccusations of betrayal, and a few Ogoni chiefs were specifically pointed out by angry Ogoni youths as collaborators who were working with Shell. Added to this was the divisive influence that Dr. Garrick Leton and his group represented. Some MOSOP activists said they had evidence that the former MOSOP president had been compromised by Shell, and they rebuffed attempts by a handful of individuals to mediate between Leton and Saro-Wiwa, with a view to giving Leton his old job as president. Leton, in turn, accused Saro-Wiwa of using the youths who belonged to NYCOP to gain control of MOSOP.[30] There was dissension, true, but nothing so irreconcilable as to result in the senseless slaughter of Ogoni by fellow Ogoni.

Colonel Okuntimo, however, interpreted these debates and dialogues, normal in a truly democratic movement, as "division" between the Ogoni leadership, and decided it was the perfect time to strike. On May 12, 1994, he sent a memo to the military administrator, Lieutenant Colonel Dauda Komo, on the subject of law and order in Ogoni. The memo, which was marked "Restricted" twelve times, read, in part:

> Shell operations still impossible unless ruthless military operations are undertaken for smooth economic activities to commence . . . division between the elitist Ogoni leadership exists . . . intra-communal/kingdom formulae alternative as discussed to apply; wasting operations during MOSOP and other gatherings making constant military presence justifiable . . . deployment of 400 military personnel . . . wasting operations coupled with psychological tactics of displacement/wasting as noted above . . . initial disbursement of 50 million naira [$500,000] as advanced allowances to officers and men for logistics to commence operations with immediate effect as agreed . . .[31]

A meeting of the Gokana Council of Chiefs and Elders was scheduled for May 21 at the palace of the Gbenemene Gokana, a traditional ruler. However, a few days before the meeting, some Ogoni leaders, with Dr. Garrick Leton in the lead, alleged that they had heard "rumors" that plans were afoot to murder certain prominent Ogoni individuals. Leton said he didn't know where these rumored murders would take place, but he found it necessary to go to Port Harcourt to lodge complaints with the military administrator, who assured him he would "take care of the situation."[32] Ken Saro-Wiwa saw the link between Okuntimo's memo and the "strong rumors" Dr. Leton claimed he heard. He wrote:

> When Lt. Col. Komo [the military administrator] assured Dr. Leton and others that he would take care of the situation, he knew precisely what he was saying. He knew he had approved Lt. Col. Okuntimo's proposals, he also knew that the Gokana people would be holding a meeting at Giokoo the following day, and he knew that an election period was as good a time as any to conduct "wasting operations." All these factors were painfully lost on Dr. Leton and his colleagues.[33]

On the day of the chiefs' meeting, Giokoo and environs were bristling with government security operatives. Yet when a mob emerged seemingly out of nowhere and descended on the venue of the meeting, murdering four of the chiefs—Edward Kobani, a former commissioner in the Rivers State government who had resigned as vice president of MOSOP along with Dr. Leton; Chief Samuel Orage, another former commissioner; Chief Theophilus Orage, former secretary of the Gokana Council of Chiefs; and Albert Badey, a former secretary to the Rivers State government—not a single policeman or soldier showed up to intervene until well after the murderers had completed their task and made good their escape. Significantly, a prominent Ogoni community leader, Chief Kemte Giadom, had drawn the attention of law enforcement agents in the area to what was happening in Giokoo, but still they did not arrive at the scene until it was too late to stop the murderers.[34] Dr. Owens Wiwa, a member of MOSOP's Steering Committee, also hurried to the Bori police station to alert the officers there to the situation in Giokoo. He too was ignored.

The first part of the "wasting operations" that Paul Okuntimo had called for in his May 12 memo now accomplished, he and his men fanned out into Ogoni for their real target. Hundreds of unarmed Ogoni men, women, and children were either massacred in their homes as they slept or were driven into the bush. Ken Saro-Wiwa, who was nowhere near the vicinity of Giokoo when the murders took place, was arrested that same evening.[35] The next day, Lieutenant Colonel Dauda Komo called a press conference and declared that MOSOP was responsible for the Giokoo murders and that he had arrested "those we wanted to arrest."[36] Komo also said on national television, broadcast on May 22, that he had directed that all those responsible for the acts (meaning MOSOP leaders) be arrested, and that state security operatives were doing just that even as he spoke. Instructively, not a single investigation had been conducted into the incidents before Komo passed his guilty verdict on Ken Saro-Wiwa and MOSOP.

From then on it was open season for Colonel Okuntimo. He let his task force loose in Ogoniland, where it spread terror, rape, torture, and death, and turned thousands into refugees. While this carefully planned slaughter was going on, the military junta cordoned off Ogoni from prying journalists so the rest of the country did not realize what was going on in the Niger Delta. It took the courage and determination of Professor Ake, an internationally acclaimed scholar and Nigeria National Merit Award winner, who sent an urgent press release from his office in Port Harcourt, for Nigerians to become aware of the tragic drama unfolding in Ogoni. Professor Ake, in his press statement, which was subsequently published in several leading independent newspapers and magazines, said: "I have followed the cynical, politically motivated assumptions of culpability, the mass arrests, the detentions and blatant abuse of the law, of the person and rights of suspects and our national honor. I have checked out the beatings. Ken Saro-Wiwa, like other suspects, was severely beaten and injured even before interrogation, and his legs were chained together for ten days . . ."[37] A few days later, on July 18, the respected Nigerian daily *The Guardian* published an editorial commentary alerting the nation to what the military junta was doing in Ogoni: "For several weeks now exceedingly perturbing reports have been coming out of Ogoniland. Tales of a military siege, tales of uniformed persons rampaging at night in the villages behind the veil of a deliberate news blackout and in the confidence that the nation has far bigger worries at the moment."[38]

The authorities did not move to stop Okuntimo, however. His mercenary team fanned out into all the 126 villages, hamlets, and towns in Ogoni. Farms were destroyed, markets were regularly raided, and school buildings burned down. In all, thirty villages were reduced to rubble. Colonel Okuntimo was true to his word. He had "sanitized" Ogoni. He was later to give a chilling and graphic account of his activities in Ogoni after the Giokoo incident at a press conference in Port Harcourt broadcast by the Nigerian Television Authority:

> The first three days of the operations, I operated in the night. Nobody knew where I was coming from. What I will just do is that I will just take some detachments of soldiers, they will just stay at four corners of the town. They . . . have automatic rifle[s] that sounded death. If you hear the sound you will just freeze. And then I will equally now choose about twenty soldiers and give them . . . grenades . . . explosives . . . very hard ones. So we shall surround the town at night . . . The machine gun with five hundred rounds will open up. When four or five like that open up and then we are throwing grenades and they are making "eekpuwaa!" what do you think people are going to do? And we have already put roadblock[s] on the main road, we don't want anybody to start running . . . so the option we made was that we should drive all these boys, all these people, into the bush with nothing except the pant[s] and the wrapper they are using that night.[39]

On July 7, Lieutenant Colonel Okuntimo raided the Ogoni village of Botem Tai with a detachment of eighty soldiers. Officials of the Civil Liberties Organization, the respected Nigerian human rights pressure group, chronicled the tragic event:

> Then without warning, a running explosion of gunshots, angry bullets flying through the air and motor engines tearing through the tracks, rent the peace of the night. It was bedlam as the mourners, men, women and children sleeping on their laps, dived in various directions for cover. "They" had come, and only a foolish man would wait for them because not even the dead earn their respect.
>
> Within minutes, about 80 soldiers in uniform had arrived, riding in three army trucks and singing like American cowboys going against helpless Indians in the movies. A pathetic symphony of wailing women and children and cries of men in pain mixed with the angry voices of sadistic soldiers as military men chased people into their houses and the surrounding bushes, cursing, kicking, and shooting and slamming people with the butts of their guns. As they did so, they extorted money and valuables, assaulting and raping women and young girls, trampling on children . . . One man heaved a sigh of relief as they drove back to the military cantonment. He was Major Paul Okuntimo. He led the team.[40]

Special detention centers were opened in Kpor and Bori and were subsequently filled with thousands of hapless Ogoni. Okuntimo and his men raped as many women as took their fancy. Villages were forced to pay "protection" money to the commander, and even then they still did not escape his wrath. While Ken Saro-Wiwa, who had a heart condition, and scores of other MOSOP activists pined away in detention, suffering physical torture and deliberate starvation, the military junta was preparing the ground for their mass hanging.

When word leaked out to MOSOP activists who had been driven underground by Okuntimo about the plot, they rallied and delegated Dr. Owens Wiwa to initiate a dialogue with Shell officials. MOSOP, a nonviolent organization, had always indicated its willingness to dialogue with Shell on ways and means of securing ecological and social justice for the Ogoni. Shell, however, preferred to dismiss MOSOP activists as upstarts who did not have the support of the majority of the people. When it became obvious, however, that MOSOP was indeed the sole and legitimate platform of Ogoni aspirations, the oil company changed tack and began to claim that the movement was militant and violent and that its hidden agenda was political in nature (by which they meant secession from Nigeria).[41] All the strands of the evidence are not yet in, but the false security reports that found their way to Abuja beginning from the Ogoni Day celebrations on January 4 1993, all alluding to plans by the "militant" segment of MOSOP to produce an Ogoni flag and national anthem, were part of a grand plan to force the hands of the junta into quelling another "Biafra" uprising. And one does not need to go far to see who was orchestrating these false reports.

In three secret meetings that Dr. Wiwa had with Brian Anderson, chief executive of Shell Nigeria at the time, in the latter's Lagos home in May through July 1995, Anderson insisted that the only condition for his "intercession" with the head of the junta, General Sani Abacha, to set Ken Saro-Wiwa and the other MOSOP activists free was that MOSOP should call off the local and international campaign highlighting Shell and the junta's activities in Ogoni. According to Dr. Wiwa, Anderson also requested that MOSOP put out a press release stating that there was no environmental devastation in Ogoni. Said Owens Wiwa, "Each time I asked him to help get my brother and the others out, he said he would be able to help us get Ken freed if we stopped the protest campaign abroad. I was very shocked. Even if I had wanted to, I didn't have the power to control the international environmental protests."[42]

Officials of Shell in London later admitted that, indeed, these private meetings took place between Anderson and Dr. Wiwa, but claimed it was part of "quiet diplomacy" to resolve the Ogoni crisis.

When details of Shell's "proposal" were communicated to Ken Saro-Wiwa in his detention cell in Port Harcourt, he dismissed it out of hand, insisting that nothing short of meeting *all* the demands of the Ogoni people would do. In November, General Abacha set up a Civil Disturbances Tribunal, consisting of two government-appointed judges and a military officer, to try cases arising from the Giokoo incident. In the dock were Ken Saro-Wiwa and Ledum Mitee, president and deputy president of MOSOP, respectively. Charged along with them were thirteen others. After a fundamentally flawed and unfair trial in which evidence was given that Shell had bribed two principal prosecution witnesses to testify against Ken Saro-Wiwa, nine of the defendants were found guilty and sentenced to death on October 31, 1995, even though their lawyers had pulled out of the case, alleging bias on the part of the tribunal.[43] Shell has consistently denied the allegations that its officials bribed the two prosecution witnesses, but the company has not to date produced credible evidence to substantiate its denial.

That same day, Shell issued a press statement reminding the world that the company withdrew from the Ogoni area "in January 1993 because it was no longer safe for staff and contractors to work there in the face of growing intimidation and physical violence from members of the communities," that Saro-Wiwa was accused of a criminal offense within the Nigerian legal system, and that MOSOP was a violent organization.[44] On the morning of November 10, 1995, ten days after Shell issued this statement in London, Ken Saro-Wiwa, Barinem Kiobel, John Kpuinem, Baribo Bera, Felix Nwate, Paul Levura, Saturday Dbee, Nordu Eawo, and Daniel Gbokoo were hanged in Port Harcourt Prison. It was a Friday morning. The dew was still fresh on the grass.

Shell and Okuntimo: The Devil Finds Work

Shell, after denying for three years what everyone in Ogoniland knew for a fact, finally admitted in December 1996 that it had invited the Nigerian authorities to help put down the "disturbance" in its Ogoni concession area in at least two instances. As Andrew Rowell, the British writer and environmentalist who has written extensively on the Ogoni saga, rightly noted, Ogoni demonstrators were killed in both instances.[45] While Shell officials still strenuously protest that the soldiers they paid were not responsible for thousands of Ogoni who were either killed or maimed beginning with the Andoni attack in June 1993, Lieutenant Colonel Paul Okuntimo, commander of the Rivers State Internal Security Task Force, and a self-confessed multiple killer, has alleged that he was paid by the company to "sanitize" Ogoni and facilitate its return to its five oil fields in the area.

In the course of a conversation with the environmentalists Nick Ashton-Jones, Oronto Douglas, and Uche Onyeagucha, whom Okuntimo had earlier brutalized for visiting Ledum Mitee, one of his many MOSOP detainees, on June 25, 1994, Okuntimo revealed that he had been risking his life and those of his soldiers to protect Shell oil installations, and that he was angry with the company for not paying him as it used to.[46] Interviewed by journalists from the *Sunday Times* of London in December 1995, one month after Ken Saro-Wiwa and his eight compatriots were hanged, Okuntimo also admitted that he regularly received payments from Shell officials while he was in charge of crushing the Ogoni protests against the company. Said Okuntimo, "Shell contributed to the logistics through financial support. To do this, we needed resources, and Shell provided these."[47] Added Mitee, who was the only MOSOP member tried along with Saro-Wiwa and the others to be acquitted: "He [Okuntimo] admitted he was being paid by Shell. He said he was angry because they were no longer paying as much for the upkeep of his boys. He felt they were not grateful enough." Mitee also said he was aware that Shell rewarded Okuntimo personally. Although Okuntimo later denied the statement he made to the *Sunday Times,* apparently under pressure from his superiors in Port Harcourt, Human Rights Watch was able to establish that all through the Ogoni crisis Shell Nigeria representatives met regularly with the commander of the Rivers State Internal Security Force.[48] The managing director of Shell Nigeria at the time, Mr. Brian Anderson, did not expressly deny his company's involvement with Okuntimo. When the matter of Colonel Okuntimo and his involvement in the Ogoni massacres was raised by the *Sunday Times* journalists, Anderson replied, "I'd like to know if we were involved with somebody like that so we could stamp it out." A spokesman for Shell in London, however, said Shell Nigeria did not authorize any financial support to the military. But this statement flies in the face of the evidence provided by Human Rights Watch and Nick Ashton-Jones.

"A Fairly Brutal Person"

"From what I hear of his recent past, he is a fairly brutal person." This is how Shell's Brian Anderson described Colonel Okuntimo to the *Sunday Times* team after his services had been dispensed with in the wake of the international outcry that greeted the murder of the nine Ogoni activists.[49] "Fairly brutal" is, however, an understatement, as any individual who has had the singular misfortune of crossing the path of this rapist and psychopath will testify.

Before Lieutenant Colonel Paul Okuntimo was redeployed by the new military administrator of Rivers State to head the Internal Security Task Force, he was second in command of the Second Amphibious Brigade, which was part of the Bori Military Camp. While still a major at the camp in 1993, he was approached to coordinate the attacks on the Ogoni from Andoni territory, cleverly disguised as a "communal clash." Okuntimo also led the Korokoro expedition. Oronto Douglas narrated how Okuntimo swore to avenge himself on Ken Saro-Wiwa and MOSOP after determined but peaceful villagers foiled Shell's attempt to return to the Ogoni area: "He told us how he nearly got killed at Korokoro last year and that he will never forgive Ken Saro-Wiwa for that. 'I ordered him to be taken to an unknown place and be chained legs and hands and not to be given food. Ken will never see the light,' he asserted vehemently."[50] The British environmentalist Nick Ashton-Jones, who was present when Okuntimo made this threat, later confirmed Oronto Douglas's account of the incident in a letter to Michael Birnbaum QC, a senior English criminal lawyer who was asked by the London-based International Center Against Censorship to attend the trial proceedings involving Ken Saro-Wiwa and the other MOSOP activists.

A good number of the members of the Rivers State Internal Security Task Force that the administrator, Lieutenant Colonel Dauda Komo, set up in January 1994 were former members of the controversial National Guard, which General Ibrahim Babangida established in the last days of his imperial presidency to help perpetuate his rule. The National Guard was, therefore, the political arm of the armed forces, an army within the army, which took directives only from the Head of State. The Ogoni problem, thanks to the false security reports sent to Abuja, was a dire threat to national security and had political undertones and therefore needed the political arm of the army, which "understood" how to deal with such matters, to quickly contain it. Thus were the members of the National Guard, who had been redeployed to other formations following the dissolution of the organization by General Abacha as soon as he seized power in November 1993, given a new lease on life. Long starved of action, they descended on the unarmed Ogoni like beasts of prey.

And to command them there was Paul Okuntimo, who long before had established his reputation in the area as an unpredictable and bloodthirsty character. In Ogoni, the people have a name for Okuntimo: the Beast. And he was to live up to this epithet following the Giokoo incident in May 1994. The Nigerian Army, weakened and corrupted by decades of mediocre leadership, a process that accelerated after General Babangida and his sidekick, Sani Abacha, seized power in August 1985, has its share of villains. But Paul Okuntimo, who, significantly, was promoted to the rank of lieutenant colonel at the height of the Ogoni massacres, was the lowest of the breed. Okuntimo was neither an officer nor a gentleman. He was, and still is, a psychopath on the loose.

Other Communities, Other Massacres

On December 15, 1993, Mr. V. Oteri, who was Shell Nigeria's security adviser at the time, requested an audience with the Inspector General of Police to discuss "crucial matters relating to disruption of our operations."[51] The "matter" was permission to import half a million dollars' worth of weapons to arm the company's supernumerary police guards (Shell spy police)—at least that was what Oteri stated in his letter to the police chief. Unrest was growing in the Niger Delta communities at this time; Shell had not been able to return to its Ogoni oil fields, and company officials feared that the other oil-producing communities would follow the Ogoni example.

Nigerian law explicitly forbids commercial firms operating in the country from importing arms for their own use. But then, Shell is not just another company. Under pressure by Shell officials, who warned that "the importance of our organization on the nation's economy cannot be overemphasised," the inspector general buckled and in July 1994 gave approval for Shell to buy weapons manufactured abroad via a third party.[52] The *Observer of London,* which obtained a copy of a materials requisition form submitted by Shell to the Inspector General, revealed that the London firm XM Federal was the proposed supplier of the weapons, among which were 130 Beretta 9mm-caliber submachine guns, thirty pump-action shotguns, and 200,000 rounds of ammunition. Shell officials in London, who had earlier denied that the company ever made such a request to the Nigerian police chief, later changed their story when confronted with the evidence and claimed the deal fell through because the company was concerned about the growing unrest in the Niger Delta and did not want to inflame the situation.[53]

Shell had previously dealt with "troublesome" communities in the Niger Delta long before the instigated intercommunal clashes that broke out in the area beginning in 1993. When the people of Iko organized a peaceful protest in 1987 and demanded that Shell put a stop to the obnoxious practice of flaring unburned gas into their skies, the company reportedly invited antiriot policemen (alias "Kill and Go") to the village. Iko villagers claimed Shell officials provided the police team with three speedboats and showed them the direction to its Utapete flow station, where the aggrieved villagers were gathered. On arrival, the armed team, led by Divisional Police Officer J. B. Effiong, went to work, shooting and looting. Those who were not wounded were simply beaten up, and in the case of women, raped on the spot.[54]

On October 29, 1990, J. R Udofia, Shell's Divisional Manager (East), wrote to the Rivers State Commissioner of Police, specifically requesting that antiriot police (the same kind that terrorized the people of Iko three years before) be sent to protect the company's facilities against an "impending attack" by youths in the Etche village of Umuechem. According to Udofia, he had reason to believe that the youths were planning a violent demonstration against Shell for the following day, October 30.

In truth, no "violent" demonstration was being planned—just a peaceful march to Shell's flow station in the area. Udofia's letter to the Commissioner of Police was headed: "Threat of Disruption of Our Operations at Umuechem by Members of the Umuechem Community." Said Udofia in the letter, "In anticipation of the above threat, we request that you urgently provide us with security protection (preferably Mobile Police Force) at this location."[55] On the morning of October 30, village youths gathered at the company's installations carrying placards and singing songs. It was a peaceful and orderly affair. Not a single Shell worker was molested. After the demonstrations, Shell officials dispatched another letter, this time to the military governor of the state. A copy was also sent to the Commissioner of Police.

The next morning a contingent of Mobile Police, armed to the teeth and chanting war songs, descended on Umuechem. They did not ask questions. They opened fire on whomever they saw, exploding tear gas canisters for good measure. By midafternoon several villagers lay dead or bleeding to death from bullet wounds. Hundreds fled into the nearby bush out of fear for their lives. After chasing them for hours, the marauders went back to their base. But it was a trick. They returned just before dawn on November 1, catching most of the villagers who had returned from the bush unawares. A slaughter ensued. An estimated eighty people were murdered in cold blood, some of them as they slept. Over five hundred houses were set ablaze, and for several hours the policemen chased after domestic livestock when there were no other villagers left to kill or molest, killing goats and chickens just for the fun of it. The judicial commission of inquiry that was later set up by the government to investigate the causes of the Umuechem massacre found not a single thread of evidence of violence or threat of violence on the part of the villagers and censured the police for displaying "a reckless disregard for lives and property."[56]

The people of Bonny have for decades watched millions of barrels of oil pass through their land every day, to enrich Shell and other people while they were fobbed off with endless promises of jobs and social amenities that were never kept. Following attempts by angry youths to get Shell to give them jobs in its facilities in the area in 1992, Shell officials called in the police. In the ensuing fracas, two police vans were slightly damaged. The police returned with reinforcements, cordoned off the town, and began to shoot. Elderly citizens of the town who were too weak to flee were wounded by the flying bullets. Others were rounded up and taken to the town square, where they were beaten up and subjected to other indignities.[57]

In December 1993 the people of Nembe, tired of waiting for Shell to make good its promise to significantly improve the social and economic life of the community where it pumps out some $2.6 million worth of oil daily, decided to take matters into their own hands.[58] On December 4 the youths of the town organized a peaceful protest, took over the company's flow station, and held the workers there hostage for a brief period. On January 15, 1994, a delegation made up of Shell, OMPADEC, and government officials visited Nembe Creek to dialogue with the community leaders. Nothing came of the visit, however, and in early February a military occupation force was dispatched to Nembe.[59] The soldiers took over the town, where they harassed the local women and threatened that they would "deal" with the people of the community for daring to attack the Shell flow station. As tension heightened, on February 8 the soldiers arrested four men in the town who they claimed had stolen an air conditioner from the Shell facility in the area. They took them to the town council building and said they would be charged with economic sabotage for trying to stop Shell's oil production. According to reports, the men, one of whom was a respected chief, were forced to the ground and beaten with guns and horse whips in full view of the public. The chief said that when he tried to escape, a soldier cut his arm and the bottom of his foot with a knife. Later that evening, five hundred angry youths gathered in front of the council building and demanded that their four compatriots be set free. The soldiers refused. As the unarmed youths approached the building, intent on rescuing the four men, the soldiers opened fire on them, hitting one youth several times in the leg and another in the shoulder.

The people of Rumuobiokani, a Port Harcourt suburb, are living on borrowed time. Shell, they complain, has been encroaching on their land over the years, quietly but systematically expanding its sprawling camp toward Rumuobiokani so that it is now difficult to say where the Shell compound stops and the village begins.[60] This would not have presented any problem if the villagers had received adequate compensation for the loss of their land, 60 percent of which they say Shell is presently occupying.

The immediate cause of the confrontation between the indigenes and the company on February 21, 1994, was, however, the former's allegation that Shell deliberately labeled their land as belonging to another community, Rumuomasi, on a map of Port Harcourt published by the company. Subsequently, on December 18, 1993, the people of Rumuomasi defaced all signposts in the area bearing the name "Rumuobiokani." For Rumuobiokani indigenes, this was like salt put into an already festering wound, and in the early morning of February 21, 1994, they gathered at the Shell compound in Port Harcourt and demanded a meeting with the company's senior executives. It was a peaceful demonstration, and all the villagers wanted was a dialogue with Shell with a view to airing their long-standing grievances. Two Shell community and public relations officials, Steve Lawson-Jack and Precious Omuku, finally arrived several hours after the villagers had been singing and dancing and waving placards in front of the Shell compound. While Rumuobiokani representatives were speaking with the two officials, a contingent of soldiers, MOPO, navy and air force officers, numbering about thirty in all, arrived. They were led by Paul Okuntimo, then still a major.[61]

According to the villagers present at the scene, Okuntimo screamed at his men that they should open fire. They immediately went to work, shooting indiscriminately, throwing tear gas canisters and beating up villagers with the butts of their guns. Several people were shot and about ten people were bundled into waiting vans, which then roared away. Human Rights Watch spoke with some of the eyewitnesses in March 1995, and they gave graphic details of what transpired that morning. A twenty-five-year-old man who was shot and wounded during the February 1994 demonstration narrated his ordeal at the hands of a naval officer:

> When [the navy and police] arrived, they started using belts on elderly men. Blood was going out from their bodies. We, the young men, tried to rescue our fathers. A navy man approached me when I was trying to rescue one of our elderly men. He told me that he will shoot me to death. I was afraid. When I tried to run away . . . he shot me. He was about one meter away. He was holding two guns, one long-range, but it didn't work so he used the pistol. He shot me once in the stomach. The bullet passed through my body. One of my intestines was hanging out so I tried to push it inside with my hands. My people rushed me to the hospital to save my life. I was supposed to spend three and half months, but due to lack of funds I came home after one month. I am still having problems. Sometimes I can't trek long distances because of the pain and because I get weak.[62]

Rumuobiokani community leaders met with Shell officials again three weeks after this bloody incident, and the officials claimed that Okuntimo and his men intervened of their own volition and that Shell did not have any hand in the matter.

"Saro-Wiwa's 'Children'"

When Ken Saro-Wiwa and other Ogoni community leaders initiated the idea to launch the Movement for the Survival of the Ogoni People in 1990, they wanted to set in motion a ripple effect on all the other oil-producing communities in the Niger Delta so they too could stand up with one voice and demand social and ecological justice. By 1992 this was happening, true to Ken Saro-Wiwa's prediction. In October of that year the Ijo nation established the Movement for the Survival of the Izon Ethnic Nationality (MOSIEND) and presented the Izon Peoples Charter to the country. Modeled loosely on the Ogoni Bill of Rights, the Izon Peoples Charter called for political autonomy for the Ijo-speaking people in a restructured and viable Nigerian federation, for the right of the Ijo to control their natural resources, and for adequate compensation to the nation for the ecological adversities they had suffered for four decades due to the exploration and production activities of Shell and the other oil companies. In March 1994, MOSIEND joined forces with the Ijaw National Congress, the umbrella organization of all Ijo-speaking people in the country, to further articulate their demands and pursue their objectives through nonviolent means.[63]

Elsewhere in the Niger Delta the MOSOP initiative had fired the imagination of youths in Ogbia, a community of forty-five villages, one of which is Oloibiri, where Shell first struck oil in commercial quantities in 1956. The people of the community are not happy with their present lot. Forty years and billions of dollars later, they are still trapped in the vicious cycle of poverty, and neither Shell nor the federal government has done anything tangible to redress this injustice. Said Princess Irene Amangala, daughter and granddaughter of two previous kings of Ogbia, "People call us fools because the companies took our oil for other people to enjoy and left us with nothing."[64]

One month after MOSIEND presented the Izon Peoples Charter, in November 1992, the people of Ogbia community gathered in the "historic" village of Oloibiri and launched their own charter. They demanded the immediate repeal of the various legislation giving the federal government authority over revenue allocation; payment of rents, royalties, and 50 percent of the profits from oil extracted from their land; compensation for environmental damage; greater representation in national institutions; greater employment of Ogbia indigenes in the oil industry and national agencies; a halt to gas flaring on Ogbia lands; the burying of high-pressure oil pipelines at least five feet belowground and away from residential buildings; and greater shore protection and erosion control. The Council for Ikwerre Nationality was established in 1993 to articulate the demands of the community in a new federation based on justice and fairness.[65]

Realizing there was great advantage to be derived in unity, networking, and proper coordination of their activities, the various oil communities came together with other minority ethnic groups in Rivers, Delta, Cross Rivers, Akwa Ibom, and Edo states—twenty-eight ethnic groups in all—to form the Southern Minorities Movement. Over ten thousand youths across the Niger Delta gathered in Aleibiri village in Bayelsa State on August 18 1997, and launched the Chikoko Movement, a new anti-oil-spillage pressure group.[66]

Shell is also coordinating its activities to keep the communities in line and neutralize this "undue" environmental awareness. But the communities show no sign of buckling under. They have made their choice, taken their destiny into their own hands. It is justice or death. On November 6, 1996, as the first anniversary of the murder of Ken Saro-Wiwa and the Ogoni eight drew near, over three thousand antiriot police were redeployed from other parts of Nigeria to Rivers State and distributed to various parts of the Niger Delta, including Omoku, Ahoada, Elele, and Ogoni. All policemen of Ogoni origin were also redeployed to the newly created Bayelsa State, in a move to put them in a place where they could be easily monitored. A unit of fifty crack police officers was also stationed at the Port Harcourt cemetery where the Ogoni nine were buried.[67]

MOSOP had called for quiet and peaceful observation of the anniversary of the November 10 executions, and a weeklong program of fasting, prayer and song, and night vigils. Shell and government officials were still edgy, however, and Major Obi Umahi, who had taken over from Lieutenant Colonel Paul Okuntimo as head of the Rivers State Internal Security Task Force, went on a rampage through Ogoni, threatening and warning people not to carry out any activity to commemorate the hangings. Members of the Niger Delta Human and Environmental Resources Organization (NDHERO), a pressure group, were harassed and their homes in Port Harcourt looted by security operatives. The organization's president, Robert Azibaola, was declared wanted. On Monday, November 11, soldiers of the Internal Security Task Force raided Bane, hometown of Ken Saro-Wiwa, and raped several women and tortured men they came across. The following morning they shot dead Barida Naaku, an Ogoni indigene in Port Harcourt. The harassment, torture, rape, and murder of MOSOP activists and ordinary Ogoni alike by security operatives during the Ogoni National Day celebrations has since become a yearly ritual.

Killing Kaiama

Faced with an increasingly restive Niger Delta, and the emergence of a new generation of educated youths who were now not only openly challenging corrupt chiefs and community leaders, but were also successfully mobilizing their communities to resist the oil companies, the Abacha junta established a new security outfit, Operation Salvage, in August 1997.[68] Operation Salvage was charged with the task of "taming" youths in Bayelsa State, where the bulk of Shell's installations were located. In Rivers State, yet another security task force, Operation Flush, was set up by the military administrator, who declared that he had obtained special emergency powers from Aso Rock (General Abacha's fortified headquarters in the federal capital) to "deal ruthlessly with economic saboteurs." Four months later, in December 1997, the junta, claiming it had received security reports that Bayelsa State was becoming increasingly unsafe for the oil companies, began to put together a plan to establish a naval base in the area. In April 1998 the military administrator of Delta State, another key oil-producing area, unwittingly revealed the details of the massive military operation the junta was preparing to launch in the Niger Delta when he "suggested" the establishment of a new National Coast Guard, "comprising the army, navy, air force, antiriot police, and customs, to ensure uninterrupted economic activities" in the oil-producing communities.

It was clear to environmental and political activists in the Niger Delta that the junta was preparing to launch another Ogonilike operation, this time in Bayelsa State, where the bulk of the Ijo people, Nigeria's fourth-largest ethnic group, whose land and creeks supplied over half of Shell's daily crude oil output, lived. A fully mobilized Ijo nation, pulling together and insisting that it had had enough of economic exploitation and environmental devastation, would make MOSOP's face-off with Shell in Ogoni seem like a minor problem. The military regime and the oil companies knew this. Something had to be done, and urgently.

The sudden death of General Abacha in June 1998, and the ensuing struggle among senior army officers to replace him, delayed the operation. But not for long. In November, Abacha's successor, General Abdulsalami Abubakar, ordered the deployment to the Niger Delta of battle-tested troops returning from peacekeeping duties in war-torn Sierra Leone. The navy was also put on alert, and steps were taken to equip it for a major offensive.[69] Ijo youths were fully aware of the military buildup and the ring of steel that was being put in place to encircle the Niger Delta, but that did not deter them.

On December 11, 1998, six days after a government radio station announced that the military regime had concluded plans to build a naval base in Yenogoa, the Bayelsa State capital, over five thousand youths, drawn from the five hundred communities and forty clans that make up the Ijo nation, gathered in Kaiama village, established the Ijaw Youth Council, and adopted the now historic Kaiama Declaration.[70] The Kaiama Declaration charged that the quality of Ijo life was deteriorating as a result of utter neglect, suppression, and marginalization visited on the Ijo nation by the Nigerian state and the Western oil companies; that the political crisis in Nigeria was mainly about the struggle for the control of the oil wealth of the people of the Niger Delta; and that "the unabating damage done to our fragile natural environment and to the health of our people is due in the main to uncontrolled exploration and exploitation of crude oil and natural gas which has led to numerous spills, uncontrolled gas flaring, the opening up of our forests to loggers, indiscriminate canalization, flooding, land subsidence, coastal erosion, and earth tremors."[71]

Asserting that all land and natural resources within Ijo territory belonged to the Ijo communities and were the basis of their survival, the youths declared that the IYC had ceased to recognize "all undemocratic decrees that rob our people/communities of the right to ownership and control of our lives and resources, which were enacted without our participation and consent." They demanded "the immediate withdrawal from Ijoland of all military forces of occupation and repression by the Nigerian state. Any oil company that employs the services of the armed forces of the Nigerian state to 'protect' its operations will be viewed as an enemy of the Ijo people." Stating that

Ijo youths in all the various clans comprising the Ijo nation would take the steps to implement these resolutions beginning December 30, 1998, as the first step toward reclaiming their lives, the IYC demanded that "all oil companies stop all exploration and exploitation activities in the Ijo area. We are tired of gas flaring, oil spillage, blowouts, and being labeled saboteurs and terrorists. It is a case of preparing the noose for our hanging. We reject this labeling. Hence, we advise all oil companies' staff and contractors to withdraw from Ijo territories by December 30, 1998, pending the resolution of the issue of resource ownership and control in the Ijo area of the Niger Delta."[72]

One dark night three days later, Felix Tuodolor, an environmentalist and one of the moving spirits behind the Kaima Declaration, was waylaid by three masked men on the streets of Port Harcourt. They edged him into an alley and warned him to leave the oil companies alone "or blood will flow." The men jumped into a waiting car and sped off before Tuodolor could identify them.[73] If this was meant to intimidate the youths and force them to back down, it failed abysmally. On December 28 the Ijo Youth Council unfolded plans for "Operation Climate Change," a series of activities designed to raise environmental awareness all over the Ijo nation between the first and tenth of January 1999, culminating in nonviolent direct action whose objective would be to put out the gas flares that were polluting their environment.

As a vehicle for nonviolent protest, Operation Climate Change was uniquely designed. Tapping the veins of Ijo culture, it sought to bring the pains and travails of the people to national attention through the "Ogele," a traditional Ijo dance where stories, song, and mime are deployed to chastise the erring, heal the wounds of the injured, and invoke the spirit of the ancestors to cleanse the land in a festive atmosphere of drink and merriment.[74] In truth, Operation Climate Change was not so much an attempt to provoke the authorities and the oil companies—as the junta was later to claim—as it was a festival of cleansing and regeneration. But all this was lost on a military regime that had long ago lost touch with the common people and their way of life, and preferred to see "subversion" even in something as sublime as a song urging humankind to protect the earth and her manifold bounties.

On the morning of December 30, the day marked by the Kaiama Declaration as the commencement of activities to implement its resolutions, young women and men all over the Ijo nation trooped out to the streets and village squares in the thousands to dance and sing and voice out their grievances. They were not armed. They were not violent. They did not molest anybody. In a letter dispatched to all Ijo villages and clans on December 28, the Ijo Youth Council had emphasized the need for participants in the Ogele/Operation Climate Change festivities to be peaceful, courteous, and orderly. This was obeyed to the letter.

What the youths did not know was that the military authorities, realizing that the Kaiama Declaration was not just another piece of paper and that the Ijo Youth Council was determined to implement its resolutions, moved quietly in the dead of night to deploy several thousand troops into the Ijo area of the Niger Delta on December 28 and 29. Reuters, the British news agency, reported that two warships and fifteen thousand soldiers had been sent to the area.[75] In Bomadi, a key Ijo town in the western Delta, the military administrator of Delta State, a naval captain, played ignorant and joined the revelers in the Ogele dance, telling them that the government was aware of their plight and was taking steps to address it. He did not tell them that enormous firepower had been ranged on Yenogoa, the capital of Bayelsa State and the heart of the Ijo nation.

As the youths, dressed in black and carrying lit candles as they sang, ululated and danced in the early morning of that fateful day on the streets of Yenogoa, violence and sudden death was the last thing on their minds. The Yenogoa police station was quiet as they passed, still singing and dancing. However, as they neared Creek Haven, home and headquarters of the military administrator, an army colonel, a barrage of machine-gun fire cut a bloody path through them. As the dead hit the ground and the wounded began to scream, tear gas canisters sailed into the air, suffocating them. In the ensuing melee, a second group of soldiers rushed to the gate of Creek Haven to reinforce the machine-gun team with rifles, shooting indiscriminately into the dancers, who were now scattered in all directions. As the gun smoke cleared, three youths—Amy Igbila, nineteen; Engineer Frank, twenty-eight; and Goodluck Wong, twenty-nine—lay dead in the street.[76] Thirty of the dancers, among them several youths who had serious wounds, were thrown into the back of army trucks and taken to the police station.

The others, who were lucky to escape, regrouped some distance from Creek Haven. After hurried deliberations, a group was chosen from among them to return to Creek Haven to ask for the bodies of their compatriots, and also for the release of the detained. As they set out, approaching the new sports complex in the center of Yenogoa town, Major Oputa, commander of Operation Salvage, accosted them. Oputa had with him five truckloads of soldiers, armed to the teeth. Witnesses who saw what happened and later recounted it to an Ijaw Council of Human Rights team, as well as Human Rights Watch researchers, said Oputa ordered the delegation to go back and said he would release their colleagues in detention. But as they turned to go, Oputa's soldiers opened fire, killing three more youths.[77] Again, the rest fled into the surrounding bush, but this time carrying their wounded, trailing blood, with them.

As dusk descended on the blood-spattered town, the military administrator, Lieutenant Colonel Paul Obi, declared a state of emergency, the first of its kind in the Niger Delta since the civil war ended thirty years before. A dusk-to-dawn curfew was also imposed, with "immediate effect."[78] But not even the administrator's desperate attempt to contain the anger of the local communities—as news spread of the wanton slaughter of unarmed dancers—by instructing his troops to shoot anyone found breaking the curfew at sight, could deter the now incensed and grieving youths. Bad news, it has been said, always travels faster than good. And by morning the gory details of what had transpired in front of the military administrator's home in Yenagoa was known by everyone in all the towns, villages, hamlets, and creeks in Ijoland.

As the sun rose, touching the Niger River with crimson, several thousand youths rose with it. They were unarmed. They knew that the soldiers would be waiting for them, and that they had orders to shoot to kill. But as Blessing Ajoko, a youth leader from Ogbia, one of the Ijo clans, was later to tell reporters, "We were determined to make our voices heard. We were determined to tell General Abubakar and the oil companies that they just couldn't go on messing up our land and killing us when we stood up to protest."[79] The slaughter continued in Yenagoa, Odi, Kaiama, and several other Ijo villages and towns that morning of December 31 and on into the new year. Soldiers had positioned their machine guns, armored personnel carriers, and tanks on all streets and road junctions in key Ijo towns and villages. As the youths emerged from their homes, they were met with rifle and machine-gun fire. People going about their daily business in the confines of their compounds did not escape the flying bullets either.

Gripped by fear and panic, residents of towns such as Yenagoa fled into the bush.

The soldiers pursued the fleeing youths into the bush. They spread out into the villages, killing and maiming. They stormed the Yenagoa General Hospital, dragged out the wounded, and, waving aside the protests of the doctors and nurses, murdered them in cold blood. Houses were vandalized and looted. Women, married ones and underaged girls alike, were gang-raped. When the women's organization Niger Delta Women for Justice attempted to mobilize women in the city of Port Harcourt to protest the brutalization and rape of their fellow women, they were set upon by soldiers and antiriot police. They were beaten back with gun butts, cowhide whips, water cannon, and tear gas. The pregnant among them who could not run fast enough were set upon by trained dogs. Over fifty women were stripped naked in the street by soldiers, beaten up, and frog-marched into police cells.[80]

Special treatment was, however, reserved for Kaiama, hometown of the Ijo hero Isaac Adaka Boro, a soldier and revolutionary who first drew the attention of the world to the plight of his people by attempting to create a Niger Delta Republic in 1966, and in whose honor members of the Ijo Youth Council had signed the Kaiama Declaration in the town in December 1998.[81] Following a peaceful Ogele dance in the town on December 30, there had been skirmishes between youths and the soldiers deployed to the town on December 31 and January 1. The soldiers claimed that three of their colleagues had been killed by the youths, an accusation the youths denied, asking the soldiers to produce the bodies if indeed this was true. The soldiers were unable to do so and instead opened fire, killing ten youths and dumping the corpses in the Nun River, which runs through the town. The following morning, January 2, ten truckloads of troops descended on Kaiama, firing in the air as they arrived. They were led by a one-eyed major who had recently seen action in Sierra Leone. He wore an eye patch in a macabre imitation of the Israeli war hero Moshe Dayan.

Major One Eye unleashed his men on the town. A team headed for the home of the king (Amanowei) of Kaiama, a highly revered personage. There they met the king, Sergeant Afuniama, and other members of his council of chiefs and elders who had gathered to deliberate on the bloody events of December 31. Afuniama and the other elders were taken to the motor park at gunpoint, their hands raised in the air. One of them, Lokoja Perewaire, knowing what was to come, attempted to flee into the nearby bush. He did not make it. A burst of machine-gun fire turned his head into a bloody mess. One of the elders later described the events of that day:

> I can only tell you what I saw with my own eyes. At about ten o'clock in the morning on January 2, I was visiting Chief Ajoko. While I was there I saw a crowd running toward us saying "Soldiers are coming!" We turned to go into the next room of the house to decide what to do, and as we turned, three soldiers came and called us to come out. We went out, Chief Ajoko, myself, and two others, and the soldiers told us to lie on the ground. I was kicked in the hip. The soldiers went away and then came back and said we should move with them. As we went we met Milton Pens Arizia, Moses Ogori, Nairobi Finijumo, Chief Geigie, and Aklis Ogbugu. We were all taken to the motor park. As we got there, they sat us down under the fruit tree. Others were lying down in the gutter. Chief Ajoko was by me. A soldier just came and used his knife to cut off the bottom of his ear. The soldier took it and told him he should eat it. He refused, and one other soldier told the first, "Don't do that." They brought four corpses in a wheelbarrow. In the evening they took them away.
>
> They took us into the motor park. We were sixty-seven when we went in. They put us in three groups and guarded us with soldiers till morning. There were more than one hundred soldiers. They told us to take off our shirts. For some time they told us to look up at the sun when it was very high and they beat us if we closed our eyes. They took sand and sprayed it in our eyes. They said we should do some frog jumps. For some years I have had a problem with my right leg, which does not bend properly. Up to today I now have pain in my leg because of the frog jumps. They said we should walk on our knees with our hands on our head. Then we had to lie on our back on top of broken bottles and creep along. They also had broken bottles and used them to cut us on our backs. Then they came with machetes and told us to sit on the ground and look forward. They cut me on my head, which started bleeding—my clothes I was wearing that day are still stained with blood. They were beating us all the time for just anything. Chief Sergeant Afuniama, the traditional ruler of Kaiama; T. K. Owonaro,

the deputy chief of Kaiama; Chief Tolumoye Ajoko, traditional ruler of Oloibiri; and Pereowei Presley Eguruze, the youth president of Kokokuma-Opokuma local government area, were taken outside for "special treatment." When Chief Afuniama was brought back into the park, he fell down unconscious. A soldier came and dropped a stone on his head. He released it twice, and he said "The chief is sleeping." This was in the morning. They left his body until the evening and then took it out.

About ten that evening, January 3, another group of soldiers came, and one of them said, "Have these people taken water and food?" and he fetched water for us. Up to that time we had no water. Some were drinking their urine; about four were ready to give up had water not been given to them. The following morning the Milad [military administrator] came, with the commissioner of police and the commissioner of health and education, and said we should be handed over to the police, who then took our names and addresses, and then released us. The Milad said nothing about compensation.[82]

The body of Chief Afuniama, his head bashed in and grotesquely misshapen, was found floating in the Nun River on January 4. But this was not the end of Kaiama's travails. The soldiers turned themselves into an army of occupation, looting, raping women and underaged girls, molesting old men, and hunting down the youths. By January 7, Kaiama had become a ghost town. The inhabitants—those of them who could run—had fled into the bush and swamps. The soldiers spread out to the neighboring villages and towns. In Oloibiri the soldiers burst into the king's palace and, seeing only his ill fifteen-year-old son in bed (other members of the household had fled), shot him while he slept.[83] In Yenagoa they rounded up people indiscriminately and put them in police cells, where they tortured them every morning. The wounded who sought medical attention at the hospital were chained to their beds like dangerous criminals. The mortuary in the hospital was now choking with the corpses of murdered youths. When their parents came for the bodies, they were turned away at gunpoint. The dead were later to be dumped unceremoniously into an unmarked grave.

Although the military administrator of Bayelsa State announced on January 4, 1999, that the state of emergency had been lifted, his soldiers, commanded by Major One Eye, continue to terrorize the populace. As of

June 1999 the road from Port Harcourt, the Niger Delta's chief city, to Yenagoa was still manned by tanks and armored personnel carriers. Local people who spoke to Human Rights Watch researchers of the arrival of the soldiers in December 1998 said that "the soldiers boasted that they had come to attack the youths who wanted to stop the oil companies."[84] And they did a good job of it. Human Rights Watch estimates that "possibly over two hundred" people were killed in Yenagoa, Kaiama, and nearby communities during this period.[85] But local activists say this estimate did not take into account those who fled into the swamps and never returned. Every four days or so, six months after the first shots rang out in Kaiama, a bloated corpse would break the calm surface of the Nun River. Those who fled into the swamps are now returning. The people are still counting.

While Kaiama, Yenagoa, Odi, and Oloibiri were besieged, the Ijo communities of Opia and Ikenyan elsewhere in the western Delta were given a good dose of the Chevron treatment. When youths of Ilaje community in Ondo State attempted to occupy a Chevron oil platform to protest the company's despoliation of their freshwater and fishing grounds in May 1998, company officials called in the navy. Chevron provided helicopters, which ferried heavily armed navy personnel and antiriot police to the platform.[86] They opened fire as they neared the platform, circling in the air. Two men, Jola Ogungbeje and Aroleka Irowaninu, were killed on the spot. Although Chevron officials later said that the navy men opened fire when one of the youths attempted to disarm the officers, eyewitnesses said this could not have been the case since the youths were not armed, and in any case the officers had opened fire even before the helicopter landed, making the oil company's claim that the youths attempted to disarm the officers untenable.[87]

The helicopters were pressed into service again on January 4, 1999, this time in two small Ijo communities where Chevron has installations. Local activists say there is a link between the military attack on Yenagoa and Kaiama and those in Opia and Ikenyan. According to them, Chevron took advantage of the mass deployment of troops in the Niger Delta by the junta to ferry troops to the two villages, with the aim of wiping them out to make way for its new oil pipeline, which is routed to pass right through the village of Opia.[88] The soldiers, numbering about one hundred, came in four boats, one of which was fitted with a machine gun, and in a helicopter contracted to Chevron Nigeria. Opia was the first to be attacked. Villagers said they saw a helicopter, the kind they usually saw conveying workers to Chevron's two oil wells in the village, flying over their village.[89] At first they thought nothing of it. Then the aircraft swooped down and began firing at them. Several people were hit. Those who escaped unhurt ran into the bush. The helicopter continued circling the village, firing into the mud-and-wattle huts. Then it headed for Ikenyan village. It descended to the level of the treetops and began to spray the huts with machine-gun fire. About ten huts received direct hits and went up in flames, together with their occupants.

The survivors of this attack were just emerging from the bush to take care of the wounded and bury the dead when the soldiers arrived. They came in three boats usually used by a Chevron contractor and a navy gunboat with a swiveling machine gun mounted in front. The people of Opia and Ikenyan did not stand a chance. The traditional leader of Ikenyan, Chief Bright Pablogba, was hurrying to the beachhead to speak with the soldiers when a hail of bullets lifted him off his feet and flung him to the ground. Then the soldiers invaded the two villages, shooting at everything in sight. They shot tear gas into the air. They smashed down the door of the flimsy huts with their boots, and anyone they saw was shot. Then they set the houses on fire and went about destroying property, including the fishing boats the people relied on for daily survival. Apparently, the soldiers had been told to wipe Opia and Ikenyan from the surface of the earth. They executed their orders with brutal efficiency.

A Human Rights Watch team visited the two communities five weeks later and saw a people still reeling with shock:

> When Human Rights Watch visited both communities in February 1999, the death toll was still uncertain. Only four bodies had been found, but a woman and her five children fishing from a canoe by Ikenyan village were also presumed dead, since the boat was sunk and they had not returned. Fifteen people from Opia and forty-seven from Ikenyan were still missing: those who still remained in the villages believed they were dead, and that their bodies had been thrown in the river or taken away—given the isolated position of the two communities it is unlikely that they could have simply fled without anyone knowing. In Opia, which previously had a total of perhaps fifty or sixty houses, we counted forty-six completely destroyed by fire, and others

> were damaged. In Ikenyan, about fifty homes were destroyed, and only four left standing at one end of the village. Tear-gas cannisters and cartridge cases were still scattered on the ground.[90]

Chevron's version of the story is that a group of youths from Opia and Ikenyan came to one of its rig locations on January 3 to demand money. Chevron acknowledged that its officials reported the matter to the military detachment in a nearby naval base who warned off the youths. The following day, according to Chevron, the youths returned to the rig in increased numbers and fully armed and engaged the armed forces in a shoot-out. Chevron claimed it was not aware of any casualties from these incidents, and that allegations that it facilitated the military expedition on Opia and Ikenyan had no basis in fact.[91]

A joint team composed of officials from the respected Nigerian human rights organization Civil Liberties Organization (CLO), the Environmental Rights Action/Friends of the Earth Nigeria, and the Ijo Council for Human Rights investigated the razing of the two communities. They said that Chevron was disingenuous when it claimed that the youths were armed and that they had engaged the security detail in a shoot-out. "That is ridiculous," said Patterson Ogon of the Ijo Council for Human Rights, who led the team. "These are poor fisherfolk. Where would they find the money to buy guns?"[92] Ogon also dismissed Chevron's claim that it was not aware of the deaths resulting from the invasion of the two communities on January 3. "Ikenyan and Opia were razed to ground. Many people were murdered in cold blood. The soldiers who carried out this dastardly act came from the Madagho military base near Chevron's operational base at Escravos. They came in a Chevron helicopter and boats used by Chevron contractors. News of the slaughter was widely reported in the newspapers the next day. So how can they say they didn't know about the deaths? Where do they live, in Mars? Let's face it, these people were killed and their villages razed to make way for the Chevron pipeline to pass through Opia. It's cheaper. Chevron wouldn't have to incur the extra cost of rerouting the pipeline. Dead people don't ask for compensation or insist on a proper environmental impact assessment, do they?"

Chevron officials in the United States continue to maintain the fiction that their company "has no involvement in or connection to any internal police activities in Nigeria."[93] This was after one of Chevron's senior managers in Nigeria, Olusola Omole, told the world on the Pacifica Radio Network, broadcasting from New York in October 1998, that his company had authorized the use of armed navy personnel at Ilaje, where two people were killed, and indeed provided the helicopters that carried out this military expedition against unarmed protesters.[94] During Chevron's shareholders' meeting in San Ramon, California, on April 29, 1999, Ken Derr, Chairman and Chief Executive Officer of Chevron, told a journalist from Pacifica Radio that his company would not officially demand that the Nigerian military desist from shooting protesters at Chevron sites.[95] In a letter to the grieving people of Opia and Ikenyan three weeks after the massacre, the Chevron public relations point man in Nigeria, Olusola Omole, claimed that "it has always been and continues to be our company's policy to have enduring and mutually beneficial relations with communities hosting our operations. Ikenyan and Opia, even though smaller settlements, have not been an exception to this rule."[96] Call it gallows humor. But the dead in Opia and Ikenyan, and indeed other Ijo communities like Yenagoa, Kaiama, and Oloibiri, are not laughing.

"Denouement of the Riddle of the Niger Delta"

If the intention of the Nigerian government and the oil companies was to cow the local communities by making an example of Kaiama and the other Ijo villages—as they did in Ogoni in 1993-94—they badly miscalculated. The massacres have hardened the resolve of the oil-producing communities to put an end to a system that has brought them so much pain, poverty, and death. Demonstrations, peaceful and well-coordinated, are now a daily occurrence all over the Niger Delta. And the effect is beginning to show. In the second week of July 1999, Shell Nigeria declared force majeure, claiming that it had suspended loading operations at its Forcados export terminal in the Niger Delta. Elf followed two days later, announcing that it was temporarily shutting down production at its Obagi oil field near Warri as a result of "threats of community unrest."[97] This triggered shock in the international energy market, pushing up oil prices to $18.69 per barrel, a twenty-month high.

Shell and the other oil companies blame their woes on "sabotage" by unemployed youths who, they also claim, have taken to kidnapping oil workers as a means of extorting money from them. Clearly, they did not pay heed to Ken Saro-Wiwa's prescient words before he was executed. In his submission to the kangaroo court that sentenced him to death by hanging on October 30, 1995, Saro-Wiwa predicted that "a denouement of the riddle of the Niger Delta will soon come," and called upon the Ogoni people, the peoples of the Niger Delta, and the oppressed ethnic minorities of Nigeria to stand up and fight fearlessly and peacefully for their rights.[98]

The Niger Delta is on the boil. The denouement of the riddle is upon us.

SEVEN

A Game for Spin Doctors

It is hard to ignore the stain of blood now spreading down that once-proud corporate logo.

Jonathan Porritt, Forum for the Future

Poor Little Rich Shell Just Wants to Be Loved

Forty-eight hours before Ken Saro-Wiwa and his eight compatriots were murdered by the Nigerian military junta on November 10, 1995, Shell swung its public relations machine into overdrive. Shortly after General Sani Abacha confirmed the death sentence on the nine Ogoni activists on November 8, Brian Anderson, chief executive of Shell in Nigeria at the time, issued a press statement in which he expressed "sympathy" for the families of Ken Saro-Wiwa and his codefendants, as well as the families of the four Ogoni chiefs. Anderson, however, refused to intervene on behalf of the condemned men even though he was aware that the judicial process leading to their conviction had been roundly criticized as fundamentally flawed. According to Anderson, "A large multinational company such as Shell cannot and must not interfere with the affairs of any sovereign state."[1]

The significance of Anderson's press statement lay not so much in what was said as what was carefully omitted. While a statement released only the previous week by the parent company, Shell International Petroleum Company in London, made reference to Ken Saro-Wiwa being found guilty of a "criminal offense" and stated that MOSOP was a violence-driven organization, Anderson took great pains to skirt the issue in his carefully worded three-page press statement, concentrating instead on denying allegations made in the course of the trial that his company had offered a bribe of 400 million naira (about $4 million) to certain Ogoni men to subvert MOSOP, and that Shell also masterminded the killing of the four Ogoni chiefs. The general tone of Anderson's statement was surprisingly conciliatory, calling for understanding, dialogue, and the need to listen to all points of view. Anderson did not forget to add, though, that his company had prepared a background briefing note "with full details of our response to specific allegations made against Shell at the [Ogoni Civil Disturbances] Tribunal," and that it was available for anyone who wanted a copy.[2] No one needed to be told what was going on—and indeed, what was about to happen.[3]

Still, when the first wave of the backlash triggered by the November 10 hangings hit Shell Tower in London a few days later, the multinational, which for decades had prided itself on its efficient public relations machinery, was caught napping. Shell's spin doctors knew there would be public outcry in the wake of Ken Saro-Wiwa's murder, but they had clearly underestimated the speed, scope, and ferocity of the counterattack launched by environmental and human rights groups, journalists, and other shocked and outraged individuals everywhere in the world. And the words, laced with anger and vitriol, fell thick and fast. The company was still reeling from the shock of the spate of street demonstrations and condemnation of its alleged role in the deaths of the Ogoni Nine a few weeks later when journalist Anne McElvoy called Shell's public relations department to arrange a background briefing for the piece she was writing for the *Spectator* in London on the Ogoni tragedy. All she got from an obviously confused Shell head of press was a vague "Statement of Principles," which was faxed to her. Speaking later on the phone, the official said hopefully that he didn't think she'd have any more questions after that. He was also anxious that McElvoy shouldn't "talk exclusively about Nigeria."[4]

Brian Anderson and his team of spin doctors in Nigeria had prepared elaborate briefing notes in the expectation that the postexecution furor would be mostly a local affair that could be quickly and efficiently managed. Apparently, London, The Hague, and other Western cities were not given similar priority treatment. Said McElvoy, "Never can there have been a public relations offensive as incompetent as that mounted by Shell Tower."[5] According to the journalist, Shell's PR campaign failed (at least initially) because the company was on shaky ground from the start, having polluted Ogoni and neglecting to introduce adequate cleanup measures until it was too late. McElvoy explained that the company's media counterattack was unable to make significant impact "because [Shell] foolishly treated an event which was bound to cause revulsion and demands for immediate action as if it were any other PR hiccup."

It did not take the company long to rally its forces, however. Brian Anderson had flown into London shortly after the international media exploded with anger following the murder of the Ogoni Nine. And for three weeks, at briefings for British MPs and pressure groups, he insisted that Shell had no hand in the murders or the violence that had engulfed Ogoni since 1993. Anderson also claimed that his company had called for help from the Nigerian military only once, during the "disturbance" in Umuechem in October 1990. He never mentioned Iko, Nembe, Korokoro, and the other towns and villages where soldiers and antiriot police visited terror and mayhem on defenseless people. Anderson also did not mention that Paul Okuntimo, who admitted being in his company's payroll, had planned and carried out the attacks on the Ogoni from neighboring villages and later misrepresented these as communal clashes.

Nnaemeka Achebe, Anderson's deputy at the time, and an executive director of Shell Nigeria, was to follow a week later with a whirlwind tour of Europe—the European Parliament in Strasbourg, Dublin (where he addressed the Dail Foreign Affairs Committee), and several other cities—speaking with members of Parliament, journalists, and human rights and environmental groups. Wherever Achebe and Anderson went, their speech was always the same, like a parrot trained to sing the same song over and again: Ninety percent of the net revenues from each barrel of oil go to the Nigerian junta, and Shell and the two other joint venture partners share "only" one dollar per barrel. From its "modest" profits the company spends $20 million annually on community development projects in the Niger Delta. Shell is a business organization and does not interfere in Nigeria's political affairs. Neither does the company collude with the military to perpetrate violence on the oil-producing communities. MOSOP has a political agenda (secession from Nigeria) and is merely using its ecological campaign against Shell as a leverage. There are undeniably environmental problems in the Niger Delta, but they do not amount to "devastation." In any case, 60 percent of the oil spillage in Ogoni before Shell was driven out of the area by violent villagers in January 1993 was caused by sabotage. Demands that Shell should withdraw from Nigeria are unrealistic because stopping oil production would destroy the country's economy. That Shell loves Nigeria and Nigerians, particularly the people of the Niger Delta, is beyond doubt. All poor little rich Shell wants is to be loved in return.

Plotting the Offensive

On February 15 and 16, 1993, three senior officials of Shell Nigeria, Nnaemeka Achebe, Precious Omuku, and A. Okonkwo, met with Royal Dutch Shell advisers on community relations and environment in the offices of Shell International Petroleum Company (SIPC) in Waterloo, London.[6] Two days later the three-man party moved to The Hague, where consultations were also held. The purpose of the two meetings was to provide background for discussions on strategies to be adopted in response to the Ogoni "challenge" during another meeting scheduled for February 26.

The British television station Channel 4 had screened a documentary about the plight of the Ogoni in its *Heat of the Moment* program in October 1992. The documentary generated a lot of bad publicity for Shell in Britain, the Netherlands, and as far away as Australia. But it was the groundswell of support for the Ogoni cause after the march on January 4, 1993, that finally convinced Shell officials in Nigeria that a fire was burning in their backyard and there was an urgent need to put it out before it engulfed the entire Niger Delta. The march garnered international media attention, and the attention of environmental and human rights groups.

Minutes of the February 1993 meetings in Shell offices in London and The Hague fingered Ken Saro-Wiwa as the "problem." The file notes of the London meeting read, in part: "International networking, most prominently so far involving the Ogoni tribe and Ken Saro-Wiwa, is at work and gives rise to the possibility that internationally organized protest could develop. Ken Saro-Wiwa is using his influence at a number of meetings, last year in Geneva at the U.N. Commission on Human Rights and, most recently, one organized in the Netherlands by the Unrepresented Nations and Peoples Organization. . . ."[7]

The leaked memo stated that the main thrust of the MOSOP activists was directed at highlighting the problems of the oil-producing communities by using the media and sundry pressure groups. It also noted that no matter what Shell did to improve public relations, the company would still be under pressure "until the communities feel that their case is being heard and that real benefits start to flow from the 3 percent Committee [OMPADEC]." The memo called for the public relations departments of the two offices in London and The Hague to keep each other closely informed to ensure that key players (read Ken Saro-Wiwa), what they say, and to whom, was more effectively monitored to avoid unpleasant surprises and adversely affect the reputation of the Group as a whole. It also called for quality improvements in media relations "to respond to questioning from the international press on matters that may have an impact on the Group's reputation."[8]

Shell devotes a great deal of time and money to keeping its corporate logo spotlessly clean. As Roger Moody has noted, "Shell has probably devoted a larger part of its budget, over a longer period, to selling itself to the conservationist lobby, than any other oil major. This, combined with a policy of guarded openness toward critics, has given it a clean image where it counts."[9] The company is closely linked to the prestigious Royal Geographic Society, some of whose activities it funds, and indeed prides itself on a corporate identity resembling a better-remunerated civil service. According to environmentalist Kenny Bruno, this is greenwash, "where transnational companies are preserving and expanding their markets by posing as friends of the environment and leaders in the struggle to eradicate poverty."[10]

Shell is also a past master in the art of spin-doctoring. The company retains the services of Burson-Marsteller, the largest PR company in the world and a subsidiary of the American advertising giant Young and Rubicam. Burson-Marsteller earns substantial income advising corporate clients on environmental issues, chalking up a profit of $18 million in 1993 alone.[11] The oil giant is also closely associated with E. Bruce Harrison, the sixth largest in the global PR pecking order, and also regarded as the environmentalist specialist. E. Bruce Harrison, the company's chief executive, made his name leading the counterattack following the publication of Rachel Carson's influential book *Silent Spring*, which raised awareness of environmental issues and provided the intellectual foundation for the environmental movement in the West. On that venture, Harrison worked with PR executives supplied by Shell and three other transnationals—DuPont, Dow, and Monsanto.[12]

When pressure began to mount on Shell in the late 1980s to pull out of apartheid South Africa or face a boycott of its products, it hired Pagan International, another PR leviathan, which helped the company devise an antiboycott campaign with the code name "Neptune Strategy."[13] Public figures were coopted to push Shell's position; the background of leading boycott supporters was investigated, with the hope that dirt would be dug up that could eventually be used against them; and fifth columnists were recruited to monitor and infiltrate boycott meetings. As icing on the cake, Shell formed a front group called the "Coalition on Southern Africa" to further its antiboycott campaign.

Shell also knows how to strike back when cornered, as it did in the spring of 1995 following its confrontation with Greenpeace over the Brent Spar incident. Greenpeace had run a hugely successful campaign to stop the British subsidiary of Royal Dutch Shell from dumping its redundant oil rig, the Brent Spar, into the North Atlantic. Even though British Prime Minister John Major weighed in on the side of Shell during the ensuing furor, the company, realizing there would be no PR dividend to be reaped from going ahead and dumping the rig in the face of the international outcry that followed Greenpeace's campaign, wisely decided to back down. Its chance for revenge came a little while later when Greenpeace apologized for getting a figure wrong in its "Don't Dump the Brent Spar" campaign. Launching a lethal PR counteroffensive, Christopher Fay of Shell accused Greenpeace and other environmental groups, albeit indirectly, of adopting "a single issue" perspective on an issue that called for balance, reason, and clear thinking.[14] In other words, Greenpeace was a fundamentalist group given to hysteria and muddled thinking, while such companies as Shell are sober, clearheaded, and adopt a balanced view on the environment even as they commit themselves to the desirable goal of sustainable development. Shell followed up this PR coup in February 1998 when it announced that the Brent Spar oil platform would be quartered up for use as a quayside for Norwegian ferries, adding that this was an innovative and acceptable "environmental option."[15]

In his book *Green Backlash*, Andrew Rowell has graphically documented the alliance between right-wing politicians in Europe and North America and their hatchet men, media establishments, and the big polluters in industry and government as they move to ambush the environmental movement, defang it, and turn it into another harmless "irritant." The Global Climate Coalition to which Shell belongs, and which has spent millions of dollars pushing the controversial claim that man-induced climate change is not occurring presently, is a key player in this game.[16] In Britain, antinuclear activists in the 1980s and the antiroad protesters in the 1990s bore the brunt of this counterattack, which, while it did not involve physical assaults, found expression in such labels as "communists," "scaremongers," and "ecoterrorists," making it easier to turn the state machinery against them. And yet when the Ogoni storm broke, Shell was taken completely by surprise.

For Shell, Ken Saro-Wiwa and MOSOP were a completely novel phenomenon. Consequently, it did not have a PR strategy ready to counter this new threat. Four decades of treating the people of the Niger Delta with undisguised contempt, secure in the knowledge that Nigeria's military could always clobber them into submission again when they dared protest, had taken its toll on the expertise of the company's PR machinery in Nigeria. Corruption was rife among senior company executives, and the officials manning the community relations department were inept, poorly trained, and unable to properly coordinate simple public relations activities. This "business as usual" attitude also percolated to London and The Hague, so that even though Nigeria accounted for a good slice of the Group's annual profits, the country was treated as safe-and-secure territory.

The panic that greeted the arrival of Ken Saro-Wiwa on the scene is therefore understandable. The effectiveness of the MOSOP campaign lay in its disarming simplicity. Since 1958, Shell had devastated the Ogoni environment. This clearly amounted to a double standard and ecological racism, since the company's operations in Western countries were cleaner, more efficient, and more environmentally friendly. Shell, in collaboration with the Nigerian government, had taken several billions of dollars' worth of oil out of Ogoni without giving the owners of the land adequate recompense. The Ogoni had had enough of this injustice, and so they were appealing to the world to help get Shell off their back. This was the message Ken Saro-Wiwa delivered wherever he went, and Shell had no response to it, for the simple reason that the accusations were true. No amount of "greenwash" could ever hope to launder Shell's operations in Ogoni and the other communities of the Niger Delta.

All through the "communal" clashes in 1993 and 1994, a regular flow of false security reports poured into Abuja, labeling Ken Saro-Wiwa a terrorist whose ultimate agenda was secession for the Ogoni. To a certain extent, this disinformation campaign worked. It prompted the junta in Abuja to consider the Ogoni "revolt" a high-priority issue requiring the dispatch of troops to the area. But the image of Saro-Wiwa and other MOSOP activists as bloodthirsty terrorists did not stick in international circles, where it really mattered—at least for Shell's spin doctors. And in Ogoni itself the mass of the people were still united solidly behind Ken Saro-Wiwa, despite the attempts of Shell officials to put a cat among the pigeons by offering generous financial inducements to certain community leaders to denounce MOSOP and ultimately destroy the organization from within.[17] These chiefs, whom the people contemptuously referred to as *bedele,* or vultures, had to contend with a new generation of Ogoni—young, well-educated, and determined to put an end to their collective denigration by Shell and the junta.

Ken Saro-Wiwa had been arrested and detained in May 1994 following the Giokoo incident. His popularity began to soar then, and he literally became a legend while still alive, though chained hand and foot in a dark detention cell and denied food and water. In November 1994, MOSOP and its leader were awarded the Right Livelihood Award (also known as the Alternative Nobel Peace Prize) for "exemplary and selfless courage and in striving nonviolently for the civil, economic, and environmental rights of his people."[18] Two other awards were to follow: the Goldman Award, which is the world's premier environmental prize—given to Ken Saro-Wiwa for leading the peaceful movement for the environmental rights of the Ogoni people—and the Hellman/Hammett Award of the Free Expression Project of Human Rights Watch. For Shell, these international awards translated into even more bad publicity. Ken Saro-Wiwa caged was even more dangerous than when he was free. It was one thing to be vilified for doing business with apartheid South Africa; but to be seen as devastating the environment and collaborating with a corrupt military junta in mass killings of defenseless people in order to steal their oil was the ultimate corporate-image nightmare.

On March 16, 1995, four senior officials of Shell International Petroleum Company (SIPC)—Malcolm Williams, head of regional liaison; A. J. C. Brak, group public affairs coordinator; D. Van den Broek, regional coordinator of the Western Hemisphere, and the African regional organization; and A. Detheridge, area coordinator of Nigeria and Angola—held a meeting with the Nigerian High Commissioner to Britain at the time, Alhaji Abubakar Alhaji, at Shell Centre in London.[19] Also in attendance were Colonel O. Iketubosin, the defense attaché to the High Commission, and S. A. Ekpa, the political counselor. The meeting was called to exchange information on the Ogoni issue and to work out a coordinated PR response to the campaign mounted by Anita Roddick of The Body Shop and other environmental groups, which were accusing Shell and the military junta of murdering hundreds of Ogoni in cold blood, and were calling for the immediate release of Ken Saro-Wiwa and the other MOSOP activists.

Brak, SIPC's public affairs coordinator, was particularly worried that The Body Shop, Greenpeace, Amnesty International, and various church groups, as well as an increasing number of Shell shareholders, had become involved in the campaign. For his part, Malcolm Williams was uncomfortable with the close relationship between Anita Roddick and a working partner of Ken Saro-Wiwa's caliber. When the High Commissioner suggested that an all-out countercampaign be launched, Williams called for caution, explaining that a direct attack would simply bring the matter more into the public domain and that the company was working on a more subtle strategy, holding discussions with key people in the environmental organizations, and that this was paying off.

The matter of Catma Films and the documentaries it had produced for Channel 4 Television was also discussed. After the 1992 documentary *Heat of the Moment*, Catma followed up in 1994 with *The Drilling Fields*, a harrowing account of Shell's plunder and devastation of the Ogoni human ecosystem in collaboration with the Nigerian junta. The Shell officials were worried that Catma Films wanted to produce yet another documentary on the Ogoni saga in a few weeks' time, and informed Alhaji Abubakar that their company had embarked on their own documentary film, which would present a "balanced" view of the Ogoni issue. The High Commissioner was also presented with copies of Shell's briefing notes—including background material on the Niger Delta Environmental Survey launched the previous month and specifically designed to serve as the bridgehead of a new public relations offensive.[20]

Niger Delta Environmental Survey: "I see the hand of God in this enterprise"

When Shell launched the Niger Delta Environmental Survey (NDES) with unprecedented media fanfare in Nigeria on February 2, 1995—timed to coincide with the beginning of the trial of Ken Saro-Wiwa and the other Ogoni activists by the Ogoni Civil Disturbances Tribunal in Port Harcourt—the company was walking on familiar ground. Shell's "Better Britain Campaign," a "public service" environmental awareness and protection project designed and coordinated by the PR giant Ogilvy and Mather, had been running for some two decades. But even as it posed as a "green" company, Shell was busy selling organochlorine pesticides, which have damaging and lasting effects on the environment.[21] The green mask slipped, however, when Shell was forced to postpone the 25th Anniversary of the Better Britain Awards in 1995 when it was caught up in the storm of controversy following its face-off with Greenpeace over the Brent Spar incident.[22]

NDES was a variant of the Better Britain Campaign, only this time designed to promote the image of a contrite company that had learned its lessons and was now ready to clean up its operations in the Niger Delta with the cooperation of other "stakeholders." According to Shell's briefing notes, the NDES policy objective was to:

- Recommend reform of inappropriate policies and practices which encourage social dislocation and environmental dislocation
- Address poverty-induced causes of environmental degradation and social tension
- Improve public sensitivity and understanding of environmental issues and the application of this understanding
- Strengthen the capacity of the people to identify and deal with environmental problems, in their local space and their own cultural idiom[23]

The preparatory phase, which was to cover the formation of a Steering Committee for the survey, defining its scope and "Terms of Reference" and arranging the visit of Steering Committee members to the oil-producing communities where public seminars would be held to finalize the Terms of Reference, was to take place between February and October 1995. The "Survey Phase" (November 1995–April 1996) would involve the examination of existing research work on the Niger Delta by a managing consultant to be appointed by the Steering Committee, with a view to determining the causes of environmental degradation to enable stakeholders to formulate programs to tackle them. The final phase—beginning about February 1996 and lasting twelve to eighteen months—would involve detailed fieldwork, which in turn was expected to provide the basis for various research reports and environmental action programs to remedy the environmental and socioeconomic problems in the Niger Delta.

Shell had gone to great lengths to convince the world that although it provided the initial funding for NDES, its Steering Committee was independent, and that the company would not interfere or influence the final outcome of the survey in any way. It was obvious from the outset, however, that the $2 million Niger Delta Environmental Survey was just another PR stunt. The immediate impetus for NDES, apart from the bad press generated by the Channel 4 documentaries and the media campaign led by several powerful environmental groups, was the announcement by the British company TSB that it was selling its Shell shares from its Environmental Investors Funds because of the company's environmental and social policies in Nigeria.[24] MOSOP's campaign was clearly hurting Shell, and plans for the two-year survey were quickly hammered out.

Shell carefully chose the members of the Steering Committee. Gamaliel Onosode, selected as chairman, was head of Dunlop Nigeria, a major user of Shell's products. Nnaemeka Achebe was a Shell employee. And most of the other members, particularly the Nigerians among them, were, as Claude Ake was later to point out, either government officials or connected in one way or another with the government or the oil companies. Ake himself, who was designated as the representative of the oil-producing communities, accepted the job at Ken Saro-Wiwa's insistence, in the hope that he would steer NDES in the right direction.[25] But it turned out to be one more hope misplaced—as later events were to prove.

Leaked minutes of a meeting that Shell officials had with a contractor they approached to participate in the survey showed clearly the purpose for which the multinational intended to use NDES. It was hoped that by passing off the survey as an independent initiative, the NDES "would solve the dual purpose of absolving [Shell] of all responsibility and addressing the local and international accusations that not enough is being done to mitigate environmental problems created by the oil and gas industry."[26] Shell's attitude to the local communities that NDES was supposed to benefit was also cynical in the extreme. During meetings with the communities in Port Harcourt on October 24 and 25, and in Effurun, Warri, on November 28, Gamaliel Onosode made a show of consulting with local people on the survey's Terms of Reference. This was a farcical game. But in the aftermath of Ken Saro-Wiwa's murder on November 10, and following the resignation of Professor Ake, the communities had no one to represent their interest on the Steering Committee. They were not allowed to elect or nominate another representative. In any case, Onosode declared during the Effurun meeting that the detailed field research would not be undertaken by members of the Steering Committee, but by professionals commissioned for that purpose.[27] And who would recruit and pay the "professionals"? Shell.

MOSOP, which had been excluded from the survey's Steering Committee for obvious reasons, saw through NDES as the public relations gimmick it was from the outset. MOSOP activists dismissed the survey as another attempt to hoodwink Nigerians into believing that Shell was now environmentally conscious, and called on the company to dialogue with it in order to conduct environmental and social impact assessment studies in Ogoni. Later, MOSOP's uncompromising stance found wide support when Dr. Struan Simpson of the Conservation Foundation revealed that Shell wanted a legal instrument that said the committee was jointly responsible with the company for the survey's findings. Dr. Simpson sat on the NDES Steering Committee as the representative for David Bellamy, founding director of the London-based Conservative Foundation. Realizing that Shell and NDES were not serious about fulfilling the survey's modest objectives, Dr. Simpson resigned in December 1997.[28]

The major criticism leveled against NDES is that rather than concentrating on the ecological and social impact of Shell's oil exploration and production activities, the survey instead elected to produce a catalogue of physical and biological diversity in the Niger Delta. It was pointed out that there were already over sixty reports on the Niger Delta environment in Shell's possession, which it has not made available to the public. *Delta*, an information-sharing newsletter set up in the United Kingdom in response to the execution of Ken Saro-Wiwa and his compatriots, commented, "There is already enough evidence to act: what is needed now is not more documentation but the cleaning up of the Delta, compensation to all those affected, and a fair share of the wealth that has been taken by force returned to the people. The $2 million that Shell is spending on the survey could provide clean water to all 500,000 Ogoni."[29]

In Claude Ake's letter of resignation from the NDES Steering Committee on November 15, 1995, he said, "Considering the tragic enormity of recent happenings and the crisis of conscience arising from them, NDES now seems to my mind diversionary and morally unacceptable. By all indications, what we need now is not an inventory of pollutants but to look ourselves in the face, reach down to our innermost resources and try to heal our badly damaged social and moral fabric." Ake also pointed out that NDES was "too little too late and does not represent a change of heart" on the part of Shell and the other oil companies.[30]

Professor Ake's letter came five days after Ken Saro-Wiwa and the eight other Ogoni were hanged after a flawed trial in which Shell played a key role by bribing two prosecution witnesses to testify, a fact that Shell has consistently denied but has not produced independently verifiable evidence to support its position. In its briefing notes, Shell had stated that "the ultimate success of the Survey lies in stakeholders working together creatively and harmoniously to harness the human and natural resources of the region."[31] Certainly, contributing to the turbulent atmosphere that eventually led to the judicial murder of nine of the Niger Delta's finest men was not the best way to creatively and harmoniously harness these resources. For his part, Gamaliel Onosode, a reverend, did not see the irony in his statement when he declared in the course of his meeting with the oil-producing communities in Effurun on November 28: "I see the hand of God in this enterprise."[32]

David Bellamy of the Conservation Foundation angered environmental campaigners in Britain when he refused to follow Professor Ake's example and also resign from the NDES after Ken Saro-Wiwa's hanging. Anita Roddick, Greenpeace's Lord Melchett, and Charles Secrett of Friends of the Earth accused Bellamy of being used by Shell to repair its battered image after the execution. It did not take long before cracks began to appear in the NDES wall, revealing the real purpose of the survey. A Dutch environmental consultancy had produced a two-volume report on Phase One of the project in September 1996, a report criticized by local NGOs and even some Shell Nigeria personnel for failing to state clearly what had been achieved so far and what was to be expected in the second phase of the survey. There was also a major disagreement between the Dutch consultancy and members of the NDES Steering Committee, culminating in the former pulling out of the survey. A local environmental consultancy replaced the Dutch firm, and in September 1997 brought out a four-volume document that it claimed was the final report of Phase One of the survey. David Bellamy and the Conservation Foundation withdrew their support for the project three months later. There is still talk of commencing the second phase, but nobody, least of all the oil-producing communities, are taking such talk seriously. They had predicted the way NDES would end up, and they were right.

Mandela, "Quiet Diplomacy," and Other Shell-Sponsored Fictions

Shell's spin doctors knew that international human rights groups and the green coalition would launch their own attack after the executions, but when it began, the scope and precision of the campaign was way beyond what Shell had ever encountered in the past. By November 19, Shell was beginning to rally, however, and beginning on Monday, November 20, the company launched a major PR counterattack, taking full-page ads in leading British newspapers to make a case for the LNG project, which it refused to withdraw from after the November 10 killings. Running alongside this campaign was yet another, infinitely more subtle, designed to rewrite Ken Saro-Wiwa's biography and present him as a philanderer, con man, and terrorist. The spin doctors were hard at work.

In response to calls that Shell intercede for the Ogoni Nine, Brian Anderson declared in his November 8, 1995, press statement that he was born in Nigeria, that he had spent the first twenty-four years of his life in the country, and that there were no quick-fix solutions to the "problems" in the Niger Delta. Anderson invoked the name of President Nelson Mandela and sought to convey the impression that his company was quietly working in league with that venerable statesman to work out a lasting reconciliation to the crisis in the Niger Delta. Said Anderson, "Many of those who know Africa best, like Nelson Mandela, are advocating 'quiet diplomacy' as the best way forward. We think those who currently advocate public condemnation and pressure would do well to reflect on the possible results of their actions."[33]

Apparently, the phrase "quiet diplomacy" was coined by Shell to fend off any calls for the company to intervene in the trial of the Ogoni activists.

When President Mandela began to employ the phrase in the weeks leading to the November 10 killings, he genuinely believed that working behind the scenes to exert pressure on the Nigerian military dictator, General Sani Abacha, was the best option. What he did not know, however, was that he was being used by Shell in a brilliant PR coup, designed first to ensure that Mandela did not throw his considerable moral weight behind calls for Ken Saro-Wiwa's release, at least not in public, and secondly to use his "saintly" name to endorse Shell's so-called policy of noninterference in the internal affairs of a sovereign country. The task of working on Mandela was assigned to John Drake of Shell South Africa (Pty) Ltd., and so successful were his efforts that President Mandela was still mouthing this meaningless phrase at the Commonwealth Heads of State conference in Auckland while the Nigerian junta was busy preparing the gallows on which Ken Saro-Wiwa would be hanged a few hours later.

Later, when Mandela discovered he'd been conned all along, he lashed out angrily, calling for sanctions to be imposed on Nigeria and Shell. John Drake was still on hand to explain away the "difficulties" his company had in trying to convince General Abacha to stay the executions. Drake requested an urgent meeting with the South African president on November 20, and proffered as proof of Shell's genuine intentions the copy of a letter that he claimed C. A. J. Herkstroter, head of Royal Dutch Shell, had written to Abacha appealing for clemency. Whether the letter was actually delivered was another matter altogether. Brian Anderson, managing director of Shell Nigeria, had put out a press release on November 8, 1995, some forty-eight hours before Ken Saro-Wiwa was hanged, stating categorically that his company would not appeal to General Abacha for clemency on Saro-Wiwa's behalf because "a large multinational company such as Shell cannot and must not interfere with the affairs of any sovereign state," adding that "these principles, in which we strongly believe, are embedded in Shell's Statement of General Business Principles." What could have changed so dramatically in two days to force Mr. Herkstroter to go against these "business principles" and send a letter to Abacha appealing for clemency? Drake followed up the meeting with a letter on November 24, thanking Mandela for meeting him the previous week in spite of the short notice and suggesting that Gamaliel Onosode, chairman of the "independent" NDES, be invited to South Africa to afford the president and members of his cabinet the opportunity "to hear a non-Shell view of the challenges facing the Niger Delta."[34] Drake also included in the letter a bulky briefing note on Nigeria in which Ken Wiwa, the late MOSOP leader's first son, was quoted out of context and presented as appealing to his father's murderers to seek a peaceful outcome to the country's troubles as a mark of respect to his memory!

Shell's PR campaign in the aftermath of the November 10, 1995, hangings was largely designed to tarnish Ken Saro-Wiwa's image, to convince the powerful green lobby in Europe and North America that Shell was not devastating the Niger Delta environment, and that where pollution occurs, it is the work of saboteurs like MOSOP activists. Interestingly, Union Carbide employed the same tactic in the aftermath of the Bhopal India disaster, blaming the horrific accident in its plant on sabotage. In the project of showing the world the "other side" of Ken Saro-Wiwa, Shell has found common cause with the trinity of Andrew Neil, former editor of the London *Sunday Times*; Donu Kogbara, another journalist, who is, incidentally, Saro-Wiwa's niece; and Richard D. North, the ex-*Independent* journalist whose controversial book, *Life on a Modern Planet: A Manifesto for Progress,* is a battering ram of the resurgent right-wing attack on the environmental movement in the United Kingdom. Several other journalists, including Dominic Midgley, a contributor to the London-based *Punch* magazine, also joined the Saro-Wiwa bashing campaign.[35]

Six days after the Ogoni nine were murdered, Andrew Neil wrote an article in *The Mail* of London titled "Saint Ken—Violent Hero of the Gullible," in which he implied that the late MOSOP leader was in reality a terrorist who had charmed his way into the liberal circles of green London, using Anita Roddick as a stepping-stone.[36] North has also been writing articles along the same lines, even suggesting that the MOSOP leader was a corrupt man and this was why he could afford to send his children to elite schools in England. It did not occur to him that Ken Saro-Wiwa was a successful writer, businessman, television producer, and publisher in his own right.

Donu Kogbara has chosen a different tack, the more lethal for its "objectivity." Kogbara begins her anti-Saro-Wiwa articles with effusive praise for her late uncle's literary talents, his urbane and cosmopolitan outlook, and his generosity, not forgetting to add that he had in fact bought her very first typewriter. Then she lays into him. Uncle Ken was inordinately ambitious, he was a con man who used others to get what he wanted, and then he discarded them—but above all he was a violent man who was single-handedly responsible for all the tragedy that befell her beloved Ogoni people since the January 1993 march. This, naturally, is sweet music in Shell's ears, and the company's spin doctors have ensured that Ken Saro-Wiwa's "poor little brave niece" is never in want of an audience to recount her tales. Shell's public relations department then carefully assembles these newspaper articles critical of Saro-Wiwa and mails them to people (including schoolchildren) who had expressed concern over the multinational's activities in the Niger Delta.

Shell has employed the British PR firm Shandwick to beef up its team of spin doctors.[37] The company has also been arranging visas and sponsoring journalists' trips to Nigeria to "see" things for themselves. Remarked *Delta*, "Dutch, German, British and other journalists have been visiting Shell installations and meeting Shell staff at Shell headquarters during these Shell trips. The program doesn't actually include much of a visit to the Niger Delta itself but they may see it by air; apparently it looks quite beautiful."[38] The company has also been putting a spin to its advertorials in newspapers and magazines to support its controversial claim that it spends $20 million on development projects in the Niger Delta every year. The people of Ojobu in Delta State were very angry when they saw a picture of their village in a Shell calendar in January 1997, with the company claiming that it had sandfilled a swampy area of the community as part of its assistance program. When they complained about this deliberate deception, state security operatives went after their community leaders. There is also the case of Chief Opara, whose wife Jacinta "mysteriously" appeared in a Shell advertorial in the London-based *Africa Today* magazine in May 1996. Shell passed her off as Mrs. Christina Ogbuyewe, and the twins in the photograph as hers, born in Erhoike Hospital, whose state-of-the-art facilities were supplied by—who else—Shell. But the hospital featured in the photograph is in Egbema and not anywhere near Erhoike. Opara and his wife have since sued Shell for defamation and are asking for 25 million naira (about $250,000) with which they hope to establish a fund for the welfare of the Niger Delta's exploited children.[39]

Shell presently operates some sixty websites to push its position. Its senior officials are constantly on the wing, traveling the globe and telling opinion leaders and key persons in the environmental movement that the company is in fact the victim in the Niger Delta saga, not the villain. Nick Ashton-Jones, the environmentalist and adviser to Environmental Rights Action (ERA), attended one such meeting in Sweden on September 20,

1996.[40] In attendance were two representatives of the Swedish Society for Nature Conservation (SSNC); three senior executives of Shell Sweden, including the managing director, a representative of Friends of the Earth, and Barbara Lawrence; and Nnaemeka Achebe from Shell International in London. Shell's only defense, put up by Achebe, was that he (Achebe) was a Nigerian, and so how could anyone think that as a Nigerian he did not want anything but the best for his country? As far as he was concerned, Shell in Nigeria was doing nothing wrong, and ERA and MOSOP had a political agenda. Achebe had given an interview to the *Irish Times* the previous February in which he implied the Ogoni were using his company to pursue a secessionist agenda.[41]

"Not in Our Character"

The Nigerian military junta's PR machinery, lethargic and inefficient even at the best of times, was completely wrong-footed following the November 1995 killings. General Abacha even told close aides that he did not understand what the fuss was all about over the death of just one man. On the occasion of the launch of the Sani Abacha Foundation, yet another stunt to launder the junta's image under the rubric of a "Not in Our Character" media campaign, Abacha said, "Ken Saro-Wiwa is being hailed as a hero, as a pro-democracy activist and an environmental campaigner, whereas it is no secret that 60 percent of oil spillage in Ogoniland are attributable to sabotage at his behest."[42] Exchange the stiff army general's tunic for a three-piece and it could easily have been a Shell official talking.

At the height of the military operations in Ogoni in which thousands were killed, rendered homeless, or forced into exile, the junta had hired Van Kloberg & Associates—a PR and lobbying firm that made its name laundering the image of blood-soaked military dictatorships in Haiti, Burma, El Salvador, and Iraq—to wage a disinformation war in such newspapers as the *Washington Post* and the *New York Times*.[43] The junta also dispatched its information minister to Oxford during a high-profile public gathering to discuss the crisis in the Niger Delta at the university on May 1, 1995. Working with a team of skilled PR practitioners from London, the minister distributed copies of an unsigned booklet, *Crisis in Ogoniland: How*

Saro-Wiwa Turned MOSOP into a Gestapo, and alleged that Ken Saro-Wiwa and his compatriots were terrorists financed by "certain foreign powers."[44]

General Abacha's way of pacifying a shocked international community after he ordered the November 10, 1995, hangings was to hire seven additional PR firms in the United States. He also spent an estimated £5 million in London (about $8 million) in a strident media campaign to obfuscate the real issues.[45] Some of the children of Ogoni chiefs sympathetic to the government were pressed into service, and they were flown to the United States and Great Britain at the junta's expense, where they attempted, albeit futilely, to sell the lie that Ken Saro-Wiwa gave the order for the Giokoo mob action. Newspaper space was also bought in these two countries, and in Holland, to peddle this fiction. Even the country's high commissions and embassies were put to use in the service of the "cause." When Trócaire, the Irish overseas development agency, launched a campaign to highlight Shell's atrocities in Ogoni by printing postcards, the chargé d'affaires of the Nigerian Embassy in Dublin sent a special booklet titled "Truth of the Matter" to Trócaire officials, in which he alleged that the late Saro-Wiwa had designed a national flag for the Ogoni, composed a national anthem, and had even secretly traveled to the United Nations to ask for political independence for his people.[46] In May 1996, when Shell first learned of the preprinted cards, John Barry of SIPC London had fired off a trenchant letter to Mr. J. Kilcollen, director of Trócaire, in which he accused the latter of "hypocrisy" for not dialoguing with his company before launching the Ogoni campaign.[47] As though Shell had dialogued with Ken Saro-Wiwa and MOSOP when they requested it.

Innocent, Like the Hornbill . . .

Shell, like the hornbill in the folk tale, took a long bath after Ken Saro-Wiwa was murdered and donned the coat of innocence. A week after the event *The Economist* magazine, a bastion of the British Establishment, struck out in the company's favor: "Environmental and human rights lobbyists have found a scapegoat in Shell, whose operations in Nigeria provide half the government's income. Shell is an easy target, but it is the wrong one. Whatever it has done to despoil the Niger Delta is now being put right. And the

US$3.8 billion project for liquefied natural gas agreed to this week is ecologically sound and stands to benefit Nigerians at large, not just their rulers."[48] That *The Economist* would descend from Mount Olympus to comment on such a trivial matter as the plight of a small tribe in the Niger Delta is surprising enough. The trouble here, though, is that logic and truth are the casualty, as the arguments and "facts" marshaled in the magazine's lead article simply do not square with the reality of the communities' lives in the Niger Delta. Ironically, a hundred years ago the same magazine in its June 16, 1900, edition criticized the decision of the British government to annex the Niger Delta, which it described as "a malarious swamp . . . which will cost several times the actual worth of its product."[49]

Shell is still pressing ahead in a frenzied effort to consolidate its PR successes, pitifully few though they are. On January 30, 1996, Cor Herkstroter, chairman of Shell's committee of managing directors, issued a statement "reaffirming" the company's support for the Universal Declaration of Human Rights and pointing out that Shell had recently met with representatives of Amnesty International and Pax Christi, "whose immediate concerns were for those who are currently detained by the Nigerian authorities."[50] Implicit in this curious press statement was the subtle message that Shell was working, again through "quiet diplomacy," to effect the release of nineteen other Ogoni still in General Abacha's gulag at the time. What Herkstroter did not add, though, was that a number of Shell's own private police had assaulted some of the detainees in the first place before handing them over to Lieutenant Colonel Okuntimo.[51]

The company has also been cleaning out its rotten Augean stables. First to go was the pair of Okuntimo and Steve Lawson-Jack, who clearly had outlived their usefulness. Okuntimo was conveniently reassigned to other duties by orders from Abuja following his "indiscreet" comments to foreign and local journalists that Shell had in fact been his paymaster all along as he waged his bloody war on the Ogoni. Brian Anderson suddenly discovered that Okuntimo was "a fairly brutal person" and quickly moved to distance his company from him.[52] As for Lawson-Jack, his days as head of public and governmental affairs were numbered when it became public knowledge that he had played a key role trying to influence certain Ogoni community leaders to subvert MOSOP. The image maker had become a liability himself, and company officials did not have to dig deep to rake up enough dirt with which to smother Steve Lawson-Jack.

On July 19, 1996, Shell in Nigeria announced new management appointments in its PR department, aimed at "strengthening the company's capability to defend its image in Nigeria and internationally."[53] Nnaemeka Achebe "retired" from Shell Petroleum Development Company Ltd. (SPDC) and joined Shell International London, where he now represents both the Shell Group and SPDC in the many international venues and forums presently focusing on Shell's devastation of the Niger Delta environment and its "pacification" of the inhabitants in collaboration with the Nigerian junta. B. E. Omiyi, the company's Operations Manager (West), was moved to a newly created position—General Manager Relations. Omiyi now coordinates environmental issues as well as public affairs, governmental relations, external communication, and media information across the entire company. Obviously, this is intended to remedy situations, such as those at the height of Shell's face-off with MOSOP and the international environmental movement, when senior company officials were contradicting each other in their haste to tar Ken Saro-Wiwa and the organization he led with accusations of terrorism and secession. Additional staff have also been recruited in the company's public relations department in Lagos, Warri, and Port Harcourt. It is unlikely, however, that the recent personnel changes will achieve the desired results. The trouble with telling lies and passing them off as gospel truth is that you have to tell yet bigger lies to paper over the yawning holes in the first one, and on and on in an ever-widening cycle that ultimately consumes the vendor of disinformation himself.

Shell has been caught out in so many false statements regarding the true nature of its activities in the Niger Delta that its public relations department has ceased to be a reliable source of information for journalists. Rather, the latter now rely on local NGOs and such international bodies as Human Rights Watch and the World Council of Churches, which sent a team into Ogoni in early 1996 to investigate and chronicle the injustices visited on a defenseless people by the oil company. Even when Shell claims, as it did in January 1995, that 285 miles of its old and dilapidated flow lines had been replaced since 1993, that all new pipelines would be laid underground, and that by 1998 the flow-line renewal program would be concluded, including a complete refurbishment of its seventy-nine flow stations, environmentalists and the oil-producing communities themselves merely paused to listen and continued with their campaign to bring the company to justice. Bitter experience has taught them to believe Shell only when it actually implements what it says it wants to implement. And even in this seemingly serious matter of what Shell calls "environmental improvement," the spin doctors have been at work. The impression was conveyed that the company would put aside about 20 percent of its annual expenses annually from 1995 to 1998 to undertake a far-reaching improvement of its operations in the Niger Delta to minimize adverse impacts on the environment. As the 1996 Greenpeace report has pointed out, however, "most of the money is going on replacing obsolete pipelines and facilities."[54] Nor did Shell add that the bulk of this money will be coming from the NNPC as the senior partner in the joint venture operations.

In December 1996, Shell spin doctors withdrew a major complaint about the updated version of the Channel 4 documentary *Delta Force*, broadcast on November 2, 1995, from the British Broadcasting Complaints Commission (BCC). In a lengthy and bulky submission running into hundreds of pages, the company had complained against Ken Saro-Wiwa's statement quoted in the program that "there is an alliance between oil companies such as Shell and the Nigerian regime" and that of his son, Ken Wiwa: "My father's real crime has been to organize the successful and nonviolent struggle against the exploitation of Nigeria's Ogoni people and land by the Shell Oil company and the Nigerian government." Shell's complaint included such other aspects of the documentary as the Ebubu oil spill and Shell's less than satisfactory efforts to clean up and rehabilitate the damaged environment properly. The company claimed that the area of damage associated with the spill, contrary to the documentary, had been significantly reduced, "and in view of the limited natural restoration until 1985, is testimony to the positive impact of the clean-up operation."[55] The company also dug up an elderly man who claimed he worked for Shell in Nigeria as a field engineer between 1965 and 1971 to support its claim that the Ebubu spill was the handiwork of retreating Biafran soldiers during the civil war and that Shell's decision to clean up the spill was in fact a humanitarian gesture.[56]

In response, Channel 4 had defended the documentary vigorously, and produced evidence to support its position. Said an elated John Willis, program director of Channel 4, "Shell's withdrawal of its complaints is a humiliating climb-down. This company had been prepared to use all its multinational resources to attack the journalism of Channel 4—but has had to back off because it could not disprove what we said. We view Shell's action as a vindication of our program and our program makers."[57]

Despite millions of dollars of PR money and some two thousand dead Ogoni, Shell is still nowhere near achieving what it set out to do on January 19, 1993: to return to its operations in Ogoni and thus score a big public relations victory. It's not for want of trying, though. On October 17, 1996, the Nigerian daily *The Guardian* reported that officials of the company met with representatives of the Nigerian military and between them hammered out a joint plan of action, estimated at $39 million, which would ease Shell's reentry into Ogoni.[58] Shell denied the meeting. MOSOP, however, insisted that such a meeting did indeed take place, saying that Shell, pro-junta Ogoni chiefs, and some contractors hammered out a memorandum, purportedly signed later by a cross section of the Ogoni people, inviting Shell to return to the area. MOSOP activists also claimed Shell paid 50,000 naira (about $5,000) for each signature.[59] Shell insists it did not make such payments, but the MOSOP officials stand by the statements. A phony group, Youth Association of Ogoni Oil Producing Communities (YAOPCO), was also floated by Major Obi Umahi, Okuntimo's successor as commander of the notorious Internal Security Task Force. Before General Abacha died in June 1998, members of YAOPCO were being provided with military training preparatory to their taking over the "sanitization" of Ogoni when Umahi's troops eventually withdrew.

Brian Anderson gave an interview to the Lagos-based *ThisDay* newspaper on the first anniversary of Saro-Wiwa's murder in November 1996, where he claimed that his company was now involved in a dialogue with MOSOP. Said Shell's chief executive, "We kept our doors open and held discussions with all groups in the area, including MOSOP. The discussions have taken place and all the people we have spoken to welcomed the move."[60] Shell also "donated" medicine to several clinics and health centers in Ogoni and announced that it had taken over Terabor General Hospital in Gokana, Ogoni, with a view to modernizing it. MOSOP, however, dismissed all these gestures as publicity gimmicks. Said Nwibani Nwako, secretary of MOSOP's Crisis Management Committee, in a press statement in Port Harcourt on January 6, 1997: "Shell is strongly reminded that Ogoni people who have lost a whole generation of their leaders and thousands of other unsung heroes . . . cannot be placated by the pittance of paltry projects like drug donations or renovation of existing structures."[61] So much for Greek gifts and publicity stunts.

The Shell Guide to Double-Speak

Writing to J. Kilcollen, director of the Dublin-based Trócaire, on March 6, 1995, John Barry of Shell International, London, swore that "as a fellow Catholic, I honestly believe that our staff in Nigeria are doing their best to apply the Shell Group's business ethics in very trying circumstances."[62] This controversial statement, solemnly made in the name of God, is not the only one senior Shell officials have put out in a long-running campaign to pull the wool over the eyes of the international community even as the multinational, in cahoots with the Nigerian government, continues to despoil the Niger Delta.

Here is a brief checklist of quintessential Shell double-speak:

1. Shell: "Accusations of Shell being in collusion with the Nigerian military authorities is deliberate misinformation."[63]

The real story: Precious Omuku, head of public relations in Shell Nigeria's Eastern Division, admitted to the *Sunday Times* of London in December 1995 that his company had called for military "protection" in Ogoni on April 30, 1993, when Willbros, its contractor, was laying pipelines in the area.[64] The rampaging soldiers shot a man and wounded eleven others. A mother of five, Karalolo Korgbara, was also shot, and had her left arm amputated. Lieutenant Colonel Paul Okuntimo, commander of the Rivers State Internal Security Task Force at the time, also alleged, although he later denied it, that he had all along been paid by Shell. Human Rights Watch also reported that "SPDC representatives meet regularly with the director of the Rivers State Security Service and Lt. Col. Paul Okuntimo."[65] Shell cracked in December 1996, when it finally admitted to paying the military in Nigeria, although still maintaining that these soldiers were not responsible for any deaths.[66]

2. Shell: "A large multinational company such as Shell cannot and must not interfere with the affairs of any sovereign state."[67]

The real story: During the trial of Ken Saro-Wiwa in which Shell says he was charged with a "criminal offense," the oil company hired O. C. J. Okocha, a lawyer, and paid him 550,000 naira (about $5,500) to observe the proceedings on its behalf. Why did Shell officials do this if the trial had nothing to do with their company, as they claimed? Two key prosecution witnesses, Charles Suanu Danwi and Nanyone Akpa, later said under oath that they were recruited and paid by agents of Shell and the Nigerian junta to bear false testimony against Ken Saro-Wiwa so he would be found guilty and executed.[68] Not only did Shell officials meet with the Nigerian High Commissioner in London in March 1995 to discuss strategies to adopt to stop MOSOP and Ken Saro-Wiwa, but John Barry, a senior official of Shell International, also stated a few months later that his company was in possession of documents taken from MOSOP offices in Port Harcourt during a raid in July of that year.[69] How did private documents get into the hands of a Shell official?

3. Shell: "SPDC's policy is that all activities are planned and executed to minimize environmental impact. It strives for continuous environmental improvement and, like Shell companies worldwide, operates within the Royal Dutch/Shell Group's Statement of General Business Principles and the Policy Guidelines on Health, Safety, and the Environment."[70]

The real story: J. P. Van Dessel, who resigned as SPDC's Head of Environmental Studies in December 1994, because he felt his professional and personal integrity was increasingly at stake, says this of Shell's operations in Nigeria: "Shell were not meeting their standards, they were not meeting international standards. Any Shell site I saw was polluted, any terminal I saw was polluted. It was clear to me that Shell were devastating the area."[71] Supporting Van Dessel's contention, the European Parliament also described the effects of Shell's oil exploration and production activities in Ogoni as "an environmental nightmare."[72]

4. Shell: "The Niger Delta Environmental Survey (NDES) has been established as a separate legal entity, with no connections to Shell."[73]

The real story: Apart from providing the core funding for the survey, Shell attempted to put in place a legal agreement that would make the company jointly responsible with the steering committee for the survey's findings.[74] This instrument would have provided an opening for Shell to manipulate or censor parts of the final report it deemed "sensitive" or too critical of its operations in the Niger Delta. So much for independence.

5. Shell: "The situation in the Niger Delta is complex, and there are no easy solutions."[75]

The real story: The situation in the Niger Delta is neither complex nor difficult to resolve. The Ogoni Bill of Rights, the Izon Peoples Charter, the Ogbia Charter, and the Kaiama Declaration sum up the problems and their solutions succinctly: repeal of the unjust laws ceding control of all land and mineral resources in the Niger Delta to the federal government; the granting of political autonomy to the various ethnic nationalities to enable them to control and use a fair portion of their resources for their own development; and the right to protect their environment from further degradation, including adequate reparation for ecological and economic damage already sustained.

6. Shell: "In the last few years the company has stepped up support for the communities in recognition of the lack of development and is now spending some $20 million a year in its area of operations on community development."[76]

The real story: This is how a senior European Shell executive described SPDC's "development" projects in the Niger Delta: "I would go so far as to say that we spent more money on bribes and corruption than on community development projects."[77]

7. Shell: "Secession is a word that gets people's backs up."[78]

The real story: MOSOP made it clear right from inception that the Ogoni want to remain part of a new Nigerian federation founded on justice and negotiated cooperation. This excerpt from the Ogoni Bill of Rights speaks for itself: ". . . Now, therefore, while reaffirming our wish to remain a part of the Federal Republic as follows . . ."[79]

8. Shell: "We withdrew our staff [from Ogoni] on 19th January 1993 and have not attempted to resume operations since that time."[80]

The real story: In April 1993, Willbros, a Shell contractor, forced its way into Ogoni with the help of Nigerian soldiers supplied by Shell. Shell was back again in the village of Korokoro, Ogoni, six months later, and with the aide of Colonel Okuntimo, attempted to resume operations in the area but was successfully repelled by outraged villagers.

9. Shell: "The Movement for the Survival of the Ogoni (MOSOP) espouses a peaceful approach, but violence has been a feature of the campaign, some of it directed at SPDC staff."[81]

The real story: The Ogoni have always jealously guarded their reputation in the Niger Delta as a peaceful community of farmers and fishermen. Right from the onset, MOSOP's struggle for ecological and social justice was conducted through nonviolent means. No cases of violent conduct were brought against MOSOP leaders prior to the kangaroo trials of 1995. Besides, the late MOSOP leader, Ken Saro-Wiwa, always urged his fellow Ogoni to rely on their brains rather than on mere brawn to pursue their goals. "Let no blood be spilt," he told a large gathering of Ogoni people during the rally on January 4, 1993, to mark the United Nations Year of Indigenous Peoples. Saro-Wiwa's speech, widely reported by the international press, was liberally spiced with such words and phrases as "nonviolent," "dialogue," and "peaceful protest."

While still in detention in Port Harcourt Prison in 1995, Ken Saro-Wiwa penned the now world-famous hymn to nonviolent struggle, "Ogoni Star":

> *Dance your anger and your joy*
> *dance the military guns to silence*
> *dance their dumb laws to the dump*
> *dance oppression and injustice to death*
> *dance the end of Shell's ecological war of 30 years*
> *dance my people, for we have seen tomorrow*
> *and there is an Ogoni star in the sky.*

10. Shell: "The welfare of the people of the Niger Delta affects our business. We quietly bring their problems to the attention of the government. There are opportunities through private meetings to make points. We believe long term that these offer greater potential than public posturing."[82]

The real story: Brian Anderson, former managing director of Shell Nigeria, bluntly told Dr. Owens Wiwa, a member of the MOSOP steering committee, in May 1995, that he would intercede on behalf of the detained Ken Saro-Wiwa and the other Ogoni activists on one condition: that they call off their peaceful, nonviolent campaign against Shell. When Anderson did not get what he wanted, the government set the process in motion that culminated in the judicial murder of the Ogoni Nine. Apparently, this is what Shell really means by "quiet diplomacy."

11. Shell: "Shell is now actively working with a group of Nigerians called the Nigerian Economic Summit, which provides advice to government on economic and social agenda issues."[83]

The real story: Nnaemeka Achebe of Shell International, London, publicly declared recently that "for a commercial company trying to make investments, you need a stable government. Dictatorships can give you that."[84] Vision 2010, a committee established by the late dictator General Abacha and charged with transforming Nigeria into a middle-income nation by 2010, and headed by Ernest Shonekan, a former director of Shell Nigeria, was Shell's unique way of sustaining the military junta to make its investments in the country safe from pro-democracy activists and other "undesirable elements" who might want to take power and call the company to account. Vision 2010, as Shonekan publicly admitted, was part and parcel of Shell's long-term "scenario planning" for Nigeria, one that was clearly intended to strengthen the hand of the most corrupt and vicious military junta Nigerians had ever seen. Abacha's Vision 2010 has since been scrapped by the new civilian government.

12. Shell: "The LNG project will cause a considerable reduction in the flaring of the gas which is now released during oil extraction."[85]

The real story: The amount of gas being flared by Shell in the liquefied natural gas project in Bonny has only been reduced by 20 percent, and the total amount of gas being flared in Nigeria (which leads the world in indiscriminate gas flaring, thanks to Shell) by 10 percent. Clearly, this does not amount to "considerable reduction." Besides, there are serious concerns about the project's environmental and socioeconomic impact. An environmental impact assessment carried out by the British firm SGS Silvi Consult has been deemed inadequate by several environmental NGOs.

13. Shell: "In the Ogoni area, investigations show that 69 percent of all spills between 1985 and the start of 1993 have been caused deliberately by the communities."[86]

The real story: Friends of the Earth-U.K. took Shell to task on this claim with the British Advertising Standards Authority (ASA) in 1996 and the latter ruled against Shell as follows: "[Shell] provided information from their Nigerian company (SPDC) that they believed showed that 17 of the 24 spills since SPDC staff were withdrawn from the Ogoni area in 1993 were caused by sabotage. The Authority noted the information described incidents that could have been sabotage but did not substantiate this. The Authority considered the Advertisers had not given enough information to support the claim and asked for it not to be repeated."[87]

14. Shell: "SPDC's current environmental performance should be seen in the context of Nigeria and its major social and economic circumstances which drive its development program. Companies operating in this setting are similarly affected. These realities are acknowledged in Principle 11 of the Rio Declaration."[88]

The real story: The 1996 Greenpeace report exposed the hypocrisy and warped logic inherent in the above statement. To quote: "SPDC is attempting to exonerate itself by hiding behind the conflict between the two criteria for good environmental policy as given in the codes of behavior of both the OECD and Shell itself . . . What this reasoning comes down to is this: Shell does not stick to international environmental standards in Nigeria because the Nigerian government does not apply or enforce these standards. What do these standards mean to Shell? Putting the corruption and military dictatorship in Nigeria to one side, it is possible that a developing country like Nigeria could have legitimate, or at least understandable, reasons for not giving priority to the enforcement of all international environmental standards. But are these same reasons valid for Shell as well? Is there anyone in Nigeria who would object to Shell investing some of its immense profits in better facilities? Of course not. Furthermore, Shell's own code of behavior prescribes that the company apply international environmental standards everywhere, even where they are not required by local law. In this light SPDC's appeal to Principle 11 of the Rio Declaration is rather improper."[89]

Business as Usual

Jedrzeg George Frynas, an environmentalist and energy economist at the University of St. Andrews, Scotland, paid a visit to the Niger Delta in February 1997 with a view to testing the claims of Shell officials that the company's operations in the area were now cleaner. Specifically, Frynas visited the communities of Okoroba, Nembe, Sangana, Anyama, Kolo Creek, and Ekole River to evaluate firsthand the general progress Shell had made in such areas as environmental protection, community projects, and human rights.[90]

What Jedrzeg Frynas saw shocked him. Based on his findings, Frynas wrote a monitor report for ERA. The report concluded that Shell had not improved on its environmental and human rights performance in the Niger Delta, contrary to its claims in the media, and that the company still treats environmental and social problems in the Niger Delta as purely a public relations matter. Frynas saw a confidential report written by officials of the Shell-sponsored NDES on the company's community projects, and their finding was that Shell officials, particularly those who were not from the oil-producing areas, had no sympathy for the plight of the Niger Delta communities, generally saw them as indolent, and also regarded the whole exercise (Shell's community development projects) as a waste of time.[91]

Dr. Wolfgang Mai, of the German development NGO *Bread for the World,* also got a taste of "Shell-speak" when he visited the company's operations in the Niger Delta in October 1997. According to Shell's promotional leaflet *Environment 1996*, twenty-two environmental impact assessments were completed in 1996 and "these EIAs are publicly available in Nigeria."[92] However, when Mai and his colleague arrived at Shell offices in Nigeria and asked to see these EIAs, not a single one was given to them to examine.[93]

This was six months after Shell modified its Statement of General Business Principles undertaking, to now "conduct business as responsible corporate members of the society, to observe the laws of the countries in which they operate, to express support for fundamental human rights in line with the legitimate role of business, and to give proper regard to health, safety, and environment consistent with their commitment to contribute to sustainable development."[94] Perhaps the simple matter of tendering the results of environmental impact assessments that it claimed it had conducted in its Niger Delta operations is not part of Shell's "commitment to contribute to sustainable development."

Bowing to pressure from such groups as Environmental Rights Action and the influential U.K.-based Pensions Investments Research Consultants Ltd. (PIRC), the Shell group appointed a senior manager to be responsible for corporate responsibility issues, published a report on its operations in Nigeria, and also produced a booklet on human rights issues that it claimed would be distributed to Shell managers worldwide. At its 1998 annual shareholders meeting, Shell International also launched a new report, *Profits and Principles—Does There Have to Be a Choice?*, describing "how we, the people, companies, and businesses that make up the Royal Dutch/Shell Group are striving to live up to our responsibilities—financial, social, and environmental."[95]

Some international human rights and environmental NGOs have hailed these developments as a "major" review of Shell's attitude toward the oil-producing communities of the Niger Delta, while waiting for actual performance on the ground. They are right to be cautious. What is not so widely known is that Shell's *Profits and Principles* booklet is part of a new $32 million public relations offensive, designed not only to clean up the company's image after the Ogoni debacle, but also to trigger an international debate on the "dilemma" global corporations (read Shell) face as they strive to balance the imperative of making profit with the need to respect fundamental human rights in communities where they operate.[96]

Only this is not so much a debate as a brilliant maneuver to manage the "profits and principles" "dilemma" in such a way as to present Shell as thinking through the problem in a serious and balanced way, as opposed to wild-eyed "ecoterrorists" who insist that there must be a choice between profits and principles, and that "principles" should be chosen, without even gauging the economic implications for developing countries where multinational corporations are helping to create wealth. There is a Shell-run "Profit and Principles" website on the Internet that urges individuals to "tell Shell" how to further improve on its new strategy, which is designed to "deliver the legacy of a sustainable future for the generations still to come." People are also encouraged to write to Shell's "The Profit and Principles Debate" section at Shell Centre in London. Shell-sponsored advertisements and "Responsible Business in the Global Economy" guides appear regularly in such leading British journals as *The Economist* and *The Financial Times*.[97]

One such, in the June 19, 1999, edition of *The Economist*, read: "Every business wants to make its mark. However, in the sensitive regions of the world, like our tropical forests and our oceans, the scars of industrialization are all too apparent. Our shared climate and finite natural resources concern us as never before, and there's no room for an attitude of 'It's in the middle of nowhere, so who's to know?' "[98]

Well, Otuegwe 1, the Ijo community in the Niger Delta that suffered a major Shell oil spill in June 1998, is "in the middle of nowhere." The impoverished village of Otuegwe can be reached only with a dugout canoe or small boat across treacherous swamps, then disembarking and finishing the journey on foot. Michael Fleshman, human rights coordinator of the New York–based Africa Fund, who visited Otuegwe in June 1999 to see the "new, improved" Shell at work, described his odyssey: "The spill is in extremely remote and difficult terrain and the visit itself was one of the remarkable experiences of my life . . . It was an arduous two-hour walk through the rain forest along a slippery mud track punctuated by chest-high ponds that had to be waded or, where the water was too deep, carefully (and sometimes unsuccessfully!) crossed on submerged log bridges."[99]

Fleshman made the trip at about the same time the Shell ad appeared in *The Economist*, encouraging local and global interest groups to "monitor our progress so that we can review and improve the ways in which we work." Fleshman obliged Shell, and this is what he saw, twelve months after a twenty-year-old Shell pipeline burst and laid waste to vegetation and fauna in the Otuegwe area:

> By the time we reached the site of the rupture, the fumes were so thick it became difficult to breathe. A sheet of oil covered the water in all directions, extending out into the creek and spreading throughout the region's waterways. Two men from the village, wearing only shorts, stood waist deep in the oil-soaked water, skimming the surface of the water with stained cotton rags and wringing the oil into plastic buckets. From time to time they would empty the buckets into a 55-gallon

> drum half full of crude. Later, they explained, they would carry the drum to the burn site and shovel the refuse into the flames. That the clean-up effort—if indeed that is what it is—is grossly inadequate and dangerous to the health of the workers was obvious.[100]

This is how a Shell contractor in the Niger Delta, three years after the company adopted its new Statement of General Business Principles in March 1997, is giving "proper regard to health, safety, and the environment"—by sending impoverished people, without the slightest protection, to mop up an 800,000-barrel oil spill with cotton rags, plastic buckets, and an oil drum. But then, Otuegwe 1 is, to use Shell's apt phrase, "in the middle of nowhere."

EIGHT

Healing the Wound

The time is not far off when the pollution of nature will become a sacrilege, a criminal act, even and mainly for the atheist, because of the one fact that the future of humanity is at stake.

Cheikh Anta Diop,
Civilization or Barbarism[1]

A typical Shell advertisement on British television, after green pastures and hauntingly beautiful hills have scrolled over the screen, usually ends with the rhetorical question: "Can we develop the industry we need without destroying our countryside?"

This, indeed, is the crucial question that must be posed to Shell executives by all people of conscience. They must pull together to pressure the multinational to come clean and admit that it has by its activities in the Niger Delta these past forty years subjected one of the most fragile human ecosystems in the world to appalling treatment. What is good for the Welsh valley, still teeming with rabbits and foxes and butterflies after Shell had laid an oil pipeline through it (or so the Shell television commercial assures us), should also be good enough for the Niger Delta, which provided the multinational with 14 percent of the $6.9 billion it earned in profits in 1995 alone.[2]

Environmental Rights Action has consistently argued that all ecosystems are human ecosystems, and that the human ecosystem of the Niger Delta, hitherto viable, self-regulating, and able to withstand mankind's impact upon it, is now facing collapse due to the reckless oil exploration and production activities of Shell and the other oil companies. Where once human beings in the Niger Delta were in perfect harmony with the environment, seeing themselves not just as owners of estates or oil rigs or oil fields, but custodians of the most sacred of objects—the environment and all that thrives on it—Shell and its attendant baggage—greed, profit, and modern technology—have violently disrupted this serene habitat and things have literally fallen apart.

Destroying a People's Way of Life

The human ecosystem of the Niger Delta is being strangled to death, slowly but relentlessly. With an annual population growth of 3 percent and the daily migration of young people to the few cities in search of scarce jobs in the oil industry, pressure on habitable land and the ecosystem itself has become intolerable.[3] Health indicators for the area are poorer than for the country as a whole, and the pervasive water-related diseases (malaria, dysentery, tuberculosis, typhoid, and cholera) are linked with environmental degradation, which itself is a function of the activities of the oil companies, among other factors. As Moffat and Linden observed, "Oil activity infrastructure development in the delta appears to cause more severe and extensive environmental impacts than oil pollution."[4] It must, however, be stressed that in the Niger Delta, polluted water sources—which itself is a fallout of infrastructure development—is even more of a potent threat to the communities.

Potable water is hard to find, and where it is available, it is often not treated. Housing and infrastructure are deficient. A study of housing conditions in the rural communities of Rivers and (the former) Bendel states in the early 1980s showed that only about 20 percent of the houses were habitable and that some 40 percent were in need of major repair to meet basic housing standards.[5] In Port Harcourt, housing is the most obvious demonstration of poor social infrastructure. A team of environmentalists led by Nick Ashton-Jones, which conducted a survey into living conditions in the city in December 1994 and January 1995, reported: "The housing reality for most people in the city is four to six people living in one small room, in a compound that does not have a water supply for much of the time, where each person shares a toilet with thirty to forty others and a bathroom with perhaps more, where the overall population density is 400 to 800 people per acre, and where the only open space is the street."[6]

Income is badly distributed in Port Harcourt. With the enormous oil revenue at their disposal, the oil companies, in league with senior government officials and the local business elite, secure for themselves choice residential areas, while the poor majority are banished to the sprawling waterfront slums and the other city ghettos where there is no electricity, water supply, or sanitation facilities. Here also, refuse collection and dumping is inefficient and badly managed, and waste dumps have taken over whole streets, vying with human beings for space. Consider a typical street scene: "Rubbish and smoke drift across the road; raggedly dressed rubbish scavengers, their faces masked from the fumes, appear and disappear like ghosts in this poisonous and malodorous Hades through which the cars swerve as if being drawn by the crazy nightmare; and this terrifying vision of our future is guarded by soldiers with guns, as if they are proud of it."[7]

In such "oil" cities as Warri and Port Harcourt, armed robbery, hooliganism, prostitution, and sudden, seemingly inexplicable explosions of street violence have become a way of life. Poverty and unemployment drive otherwise decent individuals to crime, and in these cities where the bulk of the citizenry are not part of the primary economic activity of oil production, and the policies of Shell and the government have made it impossible for even the leftovers of the feast to trickle down to them, crime, corruption, and police brutality are in abundance. Ashton-Jones and his team have argued that a poor urban environment and poverty are an economic cost to the society as a whole because the psychological stress that such conditions induce tend to result in inefficiency in the workplace and in the learning process, in domestic and public violence, in destruction of property and capital assets and sundry forms of antisocial behavior. Using a range of social and economic indices, they further believe that what they describe as "joy of living," and which they consider as a vital aspect of any society, is lacking in Port Harcourt: "The urban conditions of Port Harcourt conspire to destroy the joy of living, defeat hope, and to cripple human happiness. During the survey we often wondered of what value is a society and a system that results in so much human misery and degradation? The living conditions of Port Harcourt diminish the humanity of every person and every institution involved."[8] As in Port Harcourt, so too in Yenagoa, Warri, Ughelli, Bonny, and all the other urban centers in the Niger Delta.

In the rural backwaters as well, the communities are locked in a grim life-and-death struggle with Shell. Anger and discontent are sweeping through hitherto peaceful villages like brushfire in the harmattan.

Shell's failure to produce adequate environmental impact assessment studies for its operations has been well documented. But what is not so widely known is that not a single social and economic impact assessment has been conducted—at least not made public—in these communities that shoulder the burden of this multimillion-dollar industry, to quantify the extent of their loss as a result of Shell's oil exploration and extraction activities. Interestingly, the 1995 SGS environmental impact assessment report for the Shell-operated LNG project drew attention to the fact that there would be considerable adverse social and economic impacts on the communities in the vicinity when the gas plant was commissioned. The street protests that rocked Bonny, the operational base of the project, in September 1999 as work commenced is grim confirmation of this warning. Shell opened its first oil well in 1958. How many other communities have had their social and economic fabric rent asunder because of the intrusion of the multinational in their lives?

It must be conceded, though, that the environmental, social, and economic problems in the Niger Delta do not begin and end with Shell. As Nick Ashton-Jones and Oronto Douglas have shown in their baseline ecological survey of four communities in the area, the oil industry is not the only environmental threat confronting the Niger Delta ecosystem.[9] In Ogoni, for example, a vicious cycle of environmental degradation commenced in the 1960s as an expanding human farming population on a fixed portion of land began to exert more pressure on the land, shortening fallow periods and thus further degrading the environment, which in turn led to leaner harvests, harder work, and poverty and frustration. The advent of Shell in Ogoni in 1958 further accelerated and indeed worsened this vicious cycle of environmental degradation and want.

There is a direct relationship between environmental degradation and social discontent, and in Ogoni and all the other oil-producing communities in the Niger Delta, many people see Shell as the primary cause of their present problems. A 1995 study showed that canal and road construction, two activities closely associated with Shell, have precipitated some of the most extensive environmental degradation in the area.[10] Trapped between a mortally wounded ecosystem and a juggernaut that has proved impossible to tame or reform, the Niger Delta communities now see no future for themselves and their offspring. Ashton-Jones and Douglas capture the all-pervasive gloom and anomie vividly: "All members of the society appear to suffer from frustration for themselves and for their children as a result of poor agricultural yields, the lack of health, water, and education services, their apparent abandonment by the government, but above all by the manifestation of the oil industry in their midst that seems to represent huge wealth and yet has given nothing to them except for the impoverishment of their land."[11]

The Niger Delta is a dying ecosystem. Subsidence is occurring in the area due to the construction of dams upstream, the activities of the mining companies, and the reckless use of fossil fuels elsewhere in the Northern Hemisphere. It is estimated that the gradual sinking of the Delta combined with sea level rise as a result of global warming will force 80 percent of the population to move if they are not to perish as wide strips of their habitat are washed away. Conflict over land and other economic resources is on the rise as Nigeria's feckless government, in collaboration with Shell and the other oil companies, appropriate the bulk of the wealth of the area. Population growth is so rapid that it is estimated that food production will have to double in the next twenty years if starvation is not to complete the harvest of death the environmental devastation has already triggered.[12]

Shell has inflicted and is still inflicting grievous wounds on the Niger Delta ecosystem. This must stop. Van Dessel underlined the urgency of the situation when he remarked: "Too many promises and disappointments in the past have exhausted the patience and confidence of the people and the carrying capacity of the Niger Delta ecosystem."[13] He argues that the path toward environmental improvement should be a concrete, gradual, and visible chain of relatively small-scaled projects: "The focus should be on what can be done and not what some other party should be doing or what the ideal situation should be." Van Dessel adds that encouraging Shell—which meets its daily production target of one million barrels of oil under "difficult" circumstances—to start working on environmental progress in the Niger Delta might be the most effective approach.[14]

But this proposal begs the question: How do you suddenly turn a company that has waged a relentless ecological and economic war against the inhabitants of the Niger Delta for four decades away from its bad ways? Ashton-Jones frames the question even more poignantly: "Can Shell care? Can Shell show humanity? Shell is a big multinational company; maybe the biggest on Earth. So how can a company care? How can a company have human feelings?"[15] The author and environmentalist Roger Moody supplies an unequivocal answer: Shell is certainly not about to get religion and become a moral force. It would certainly be a nice thing, says Moody, if Shell could be brought to its knees in court, stripped to its barest assets after paying out massive compensation for environmental damage, land seizure, and involvement in murder. But, cautions the environmentalist, the days when David journeyed forth to vanquish the evil Goliath are over, and he illustrates this with Union Carbide's chemical holocaust in India. Warren Anderson, the chemical giant's chief executive, has never been in court, let alone faced charges, more than fifteen years after the Bhopal disaster. Indeed, Union Carbide is still growing. Environmental law, argues Moody, should mean just that: nowhere for the criminals to hide and cook up other ways of appropriating our common inheritance for future devastation. But fighting a colossus like Shell will cost money, and as Moody points out, what is the point "tying up our hands and precious piggy banks in costly litigation, with only a handful of lawyers willing to act on a pro bono basis."[16]

The enormity of the task ahead, calling Shell to account for its crimes against the Niger Delta human ecosystem and healing the wound it has inflicted on it, becomes clearer against the backdrop of the tide that is presently turning against the environmental movement worldwide, propelled by right-wing politicians and government officials aided by powerful and amoral multinational corporations. The British environmentalist and writer Andrew Rowell has rendered a penetrating account of how "industry dollars are funding politicians, think tanks, and forming front organizations to con the public whilst companies are spending billions on perpetuating a green and caring image, and public relations companies are spending millions on antigreen PR."[17]

Rowell outlines three broad areas that have to be urgently addressed if the antienvironment monster is to be beaten back and caged. The environmental movement has to recover its roots, it has to broaden out to work closely with other groups, and it has to start putting forward solutions and a positive alternative coherent vision for the future.[18] The Indian environmentalist Vandana Shiva supports Rowell's position:

> If environmental activists try and act alone, without connecting up with movements for justice, movements for human rights, movements for

democracy, they can be contained very easily, not just by backlash, but also by polarization, by constantly making it look like the environmental interest is a secondary interest, whereas jobs and the survival interest are the primary interest. This will be exacerbated by people refusing to recognize the environmental base is also a livelihood base, and environmental issues are tied very closely to economic survival. I think what is needed now very rapidly is broad-basing of our environmental work. Building into trying to create large citizens' alliances and trying to deal with deregulated commerce, and uncontrolled power of capital.[19]

Shell's Nemesis

What Rowell and Shiva recommended is already happening in Nigeria, where a broad coalition of environmentalists, human rights and democracy workers, and activist journalists came together in the aftermath of Ken Saro-Wiwa's death in November 1995 to continue the campaign for ecological and social justice in the Niger Delta. Ken Saro-Wiwa is Shell's nemesis. The Niger Delta is stirring, and its inhabitants are asking questions, urgent questions to which they are also demanding clear answers, not just the usual Shell double-speak. Barry Hillenbrand, a journalist with the American magazine *Time*, was on hand to witness one such encounter between Shell executives and the Sapele community in November 1996, the first anniversary of the MOSOP leader's murder:

> Four fans whirled overhead, shifting the hot and humid afternoon air around the open-sided community hall with very little effect. But the assembled chiefs and elders from the Sapele-Okpe community in Nigeria's Delta Region did not seem to mind the heat. On the contrary, they were intent on generating a bit more heat of their own. Their spokesman, Vinkariks Ekariko, was addressing executives from Shell Petroleum Development Co., which pumps 930,000 bl. of crude oil a day out of the Delta, a swampy morass of southern Nigeria where the Niger River empties into the Atlantic. His tone was polite but impassioned. "You produce the oil from our land," said Ekariko, "but we get no

benefit from it. Look around. Does this look like an oil-producing community? Does this look like Saudi Arabia?"

No, there is little chance of confusing the ragged shops and hardscrabble homes of Sapele township with the glittering villas and bulging shopping malls of Riyadh. Here in the Delta, where the economics of Big Oil and small African tribes intersect, the numbers are staggering in their disproportion. Royal Dutch/Shell Group, SPDC's parent, is the most profitable company on earth. Last year it earned $6.9 billion on revenues of $109.8 billion. In Nigeria some $7 billion in oil money from Shell and other companies pours into the government treasury each year. Shell clears $220 million from the Delta. The Sapele share: next to nothing.[20]

Raising the Stakes: What the Oil-Producing Communities Are Doing

MOSOP was the first properly organized response to Shell's sundry depravities in the Niger Delta. Building on the movement's successes and also learning from its reverses, particularly in the wake of the execution of the Ogoni Nine, other oil-producing communities are linking up and pooling their resources to present a common front to the oil multinationals. And already Shell is feeling the heat. While the company claims that it lost eighteen million barrels of oil production in 1995 due to "community disturbances," the truth is that the Niger Delta is no longer Shell's docile cow, to be milked at its pleasure and convenience and then discarded. The cow is insisting that it be milked properly and that it also receive adequate recompense for what it gives. Simply put, the oil-producing communities are now demanding that Shell conduct environmental, social, and economic impact assessments before a single oil well is dug in the area. They are also insisting that the obnoxious practice of flaring gas indiscriminately and polluting the atmosphere be stopped immediately, and that the high-pressure pipelines that crisscross their farmlands and backyards be buried out of sight. Above all, they are demanding a share of the action—rent and royalties for their oil, and compensation for land and other property damaged or polluted in the course of oil exploration and production activities.

The Niger Delta communities, working through such mass-based organizations as the Chikoko Movement, are raising the stakes and are forcing a choice upon Shell: Explore and produce oil on our land according to internationally acceptable standards or contend with our wrath. Elsewhere in the world, oil extraction can be a relatively clean industry. Crude oil itself is a natural product that can be easily handled by a healthy and viable ecosystem, more so in the hot and humid Niger Delta where biological cycles are fast. Shell, however, chose the way of the European colonialists of old: Invest as little as you can possibly get away with in infrastructure and regulations that will ensure a healthy environment, and take the oil and run. The "natives" can sort out the mess later. The days of cheap oil are now effectively over, though. As Ashton-Jones has pointed out, Shell and its shareholders now have to contend with three costs as a result of the new wave of environmental and social awareness in the Niger Delta communities: collateral damage to capital assets (due to lack of maintenance of machinery), as communities demonstrating against the company's obnoxious practices make it difficult for workers in flow stations and other facilities to remain on site; downtime, forcing Shell contractors to suspend work due to civil unrest in areas of operation; and lost production, as in Ogoni and Ijoland, where the company has had to close down some wells following the mobilization of the communities by their leaders. It has been suggested that a Shell-sponsored seismic survey suspended due to community protests was costing the company some $225,000 every month as long as the "disturbance" lasted.[21] Clearly, these are costs that no company, no matter how rich, can afford in the long term.

It is now common knowledge that the bulk of Shell's pipelines and flow lines are old, rusty, and corroded, and leak oil. The standard Shell response to these leaks, before local and international protests began to build up, was to blame it on sabotage. Interestingly, though, Brian Anderson, former managing director of Shell Nigeria, disclosed during a public meeting convened by the Royal Geographic Society in London in April 1996 that his company had set aside $1 billion for a flow line replacement program that would span four years, beginning from the previous year, 1995, in which $159 million had already been spent. Shell spin doctors prefer to call this an "environmental improvement" program, but what is actually going on is that, faced with local and international pressure, the company is embarking on long-overdue maintenance that has absolutely nothing to do with Ogoni "saboteurs." Argued Nick Ashton-Jones and Oronto Douglas, "Clearly Shell is embarrassed by the bad manner in which it has managed its flow lines, but what is really scandalous is that not only have they failed to practice decent management but that also they had no intention of doing the right thing until they were pressed by public opinion first in Ogoni and then, on a worldwide scale, elsewhere."[22]

Knowledge, Information, and Empowerment

Members of Environmental Rights Action (ERA), a grassroots pressure group peacefully campaigning to change the policies of government and nongovernmental organizations in Nigeria, where those policies are likely to infringe the environmental rights of the local communities, are now working actively in the Niger Delta to enable the oil-producing communities to defend their environmental human rights against Shell and the other oil companies.[23] Relying on Article 24 of the African Charter of Human and People's Rights, and the 1992 Earth's Summit declaration that sustainable development is the most effective way of reversing poverty and environmental destruction, and that sustainable development itself can only be achieved through broad public participation, ERA activists have established Community Resource Centers (CRCs), where community members as individuals or groups can empower themselves with knowledge and use tools of communication to improve their lot.

ERA's argument is that the creation of wealth without knowledge and participation is not sustainable development because if local people remain ignorant and are unable to participate in the creation of wealth, environmental degradation will result. Significantly, this tallies with the late Professor Ake's thesis that ultimately, economic development can only be sustained if its processes are democratized so local people are participants as well as policy makers. Specifically, ERA's CRCs are designed to equip the oil-producing communities to see Shell and its activities in their proper context and then demand that things that are not being done properly by the oil company are either stopped forthwith or amended.

This is one significant step that has been taken toward healing the wound on the Niger Delta's tender side. ERA's work is not limited to Shell's activities in the Delta, though. Realizing that Agip, Elf, Chevron, Texaco, and

Mobil are as guilty as Shell in their careless attitude toward the human ecosystem of the Niger Delta, the pressure group is also focusing attention on these companies, particularly Mobil's Kwa Ibo terminal at Stubbs Creek near Eket, which is threatening to wipe out the remaining tropical rain forest on the Akwa Ibom coast. Mobil tried to hide a major oil spill at its Idoho platform on June 14, 1995, from the international community, until ERA volunteers blew the whistle. Another potential environmental nightmare that ERA has beamed its searchlights on is an aluminium plant presently being built at Ikot Abasi on the Imo River.

Global Warriors, Hidden Agendas

Despite the efforts of big business and right-wing politicians to ambush the environmental movement, there is a slow but perceptible shift in consumer attitude in Western countries toward "ethics," a process in which morality has become the latest buzzword in the marketplace. The British journalist Anne McElvoy explains: "It used to be the environment which had us changing our washing powder and buying biodegradable packaging. Now we want to be sure that we are not only buying ecological soundness but moral superiority when we shop. Multinationals are not only expected to meet high environmental and ethical standards in their own operations, they are increasingly called upon to make judgements on the politics of the countries in which they are operational, and act accordingly."[24]

Shell's role in the judicial murder of Ken Saro-Wiwa and his compatriots angered these consumer groups, and they responded by orchestrating a campaign highlighting the company's activities in the Niger Delta. In November 1994, a year before Ken Saro-Wiwa was murdered, the London-based TSB Environmental Investors Fund had removed Shell from its portfolio, a clear indication that shareholders were aware of what the multinational was up to in the Niger Delta. A year later, shortly after the Ogoni Nine had been sentenced to death, Technowax Ltd., the British candle makers, announced that they would no longer buy their £130,000 (about $200,000) a year worth of fuel and raw materials from Shell in view of its activities in Ogoni. Prominent members of the Royal Geographic Society have also been pressing for the society to review its relationship with Shell, a key financial supporter of its activities.

Ireland was the first country in the Northern Hemisphere to invite Shell to explain itself before its Parliamentary Committee on Foreign Affairs following the death of Ken Saro-Wiwa. This was the outcome of three months of intense behind-the-scenes lobbying by Trócaire, the Irish Catholic Agency for World Development, which ensured that detailed submissions on the Ogoni tragedy were made to the Joint Committee on Foreign Affairs in the aftermath of the executions of November 10, 1995. Sister Majella McCarron, of the Africa-Europe Faith and Justice Network, who lived in the Ogoni area for several years, and Shelley Braithwaite, who worked with Ken Saro-Wiwa in Ogoni, played key roles in sensitizing Trócaire staff and the international community in general to the silent genocide that was taking place in this area of the Niger Delta. Trócaire had consistently argued, and quite rightly too, that insufficient attention had been paid to Shell's role in the Ogoni massacres of 1993 and 1994, which eventually culminated in the hanging of Ken Saro-Wiwa. In January 1996, Deputy Jim O'Keefe drew up a resolution calling on the Irish government to "seek the introduction of an oil embargo." Members of the Joint Committee on Foreign Affairs and the Tanaiste (Prime Minister) at the time, Dick Spring, were solidly behind the move, and representatives of Shell were subsequently called before the Committee.[25] Trócaire also led the international campaign to enforce economic and trade sanctions against the Abacha dictatorship, and drew up a list of recommendations for action to this effect.

The argument in favor of a boycott of Nigerian oil—to persuade the government to take the issue of environmental human rights in the Niger Delta seriously, and also force Shell to mend its ways—is compelling. Dr. Tajudeen Abdul-Raheem has drawn up a list of issues centering on the sanctions campaign in a twenty-question-and-answer format, arguing that oil has never benefited the mass of the Nigerian people and that a boycott now will hurt the government more than it will hurt the masses, who in any case are already impoverished by the government's irresponsible and visionless economic policies. Said Abdul-Raheem:

> The proceeds from oil does not benefit the average Nigerian. It is a monopoly of the oil companies and the tiny oligarchy of friends, families, and hangers-on of the generals. They are not building roads, hospitals, schools, or supporting any welfare programs with the money. To deny them access to that revenue can only be of marginal consequence

> to a majority of Nigerians. A nation that produces almost two million barrels of oil a day, and yet motorists and other fuel users have to queue up for hours or even days, sometimes, to buy fuel, is proof that the oil industry does not work in the people's interests. Therefore to impose sanctions on its foreign sale may even make it available at home for people to buy, and also deny the dictators their illicit wealth from the collective resource.[26]

Trócaire, like Dr. Abdul-Raheem, has also pointed out that Nigeria's per capita income has declined from $1,000 in 1983 to an all-time low of $250 in 1994, in spite of oil production, and so a boycott can only hit Shell and the government, which are raking in all the money while the overwhelming majority of 120 million Nigerians are living in poverty. Besides, Nigerian oil only amounts to one-thirtieth of total world production, and as Trócaire has argued, "an embargo on Nigerian oil would hardly have an adverse impact on global supplies or prices. Its real impact would be in disabling the current regime from continuing in power."[27]

It is, however, unlikely that the United States and the other Western countries would lead the boycott of Nigerian oil. Nigerian oil is particularly important to the United States and Western Europe, which between them imported 92 percent of Nigeria's total oil production in 1991.[28] Nigeria is the largest producer of sweet (almost sulfur-free) crude among the OPEC countries, and its oil is very popular with refineries in the United States and Western Europe, where pollution legislation is very stringent. Nigeria's oil wells are also nearer the Atlantic Basin markets, relative to Middle Eastern oil, and this, combined with the fact that the West African subregion is relatively more stable compared to the politically volatile Gulf region, yields a significant price premium and also makes the country strategically important to the West.

There is also the issue of "cheap oil," which American and European consumers, following the humbling of OPEC as a price-fixing cartel, have come to regard as a birthright.[29] A complete boycott of Nigerian oil, coming at a time the United States is leading the embargo on Libya and Iraq, two major oil producers, could send oil prices beyond their recent high, a development that would not go down well with American voters.[30] There are also the major American importers of Nigerian crude to contend with. Between them, Shell, British Petroleum, Sun, and Chevron accounted for 80 percent of Nigerian crude sold in the United States in 1990 and 1991. By 1999 the United States was importing one-twelfth of its oil from Nigeria.[31] Shell, the largest equity producer in Nigeria, also exports half of its Nigerian production to its refineries in the United States. The combined lobbying power of these multinationals, assiduously playing the strategic national interest card, likely ensures that the issue of boycotting Nigerian oil will not feature seriously in American foreign policy. Indeed, when Gilbert Chagouri, the oil trader and point man for General Abacha, "donated" the sum of $460,000 to the Democratic party in October 1996 and earned an audience with high-level officials of the Clinton administration, he was merely trying to accelerate what was already in the process: the softening of the American government's previous "hard-line" policy on Nigeria and substituting it with "constructive engagement."[32] It therefore came as no surprise when President Clinton, during his tour of African countries in March 1998, suddenly announced in Johannesburg that his administration would not object if General Sani Abacha ran for president if he did so as a "civilian."[33]

The fact that General Abacha banned independent political parties in the country and was the sole candidate for president after putting the winner of the annulled June 1993 presidential elections in detention were issues that President Clinton conveniently neglected. For the Clinton administration, what was really at stake was Nigerian oil and its uninterrupted flow to American markets, and a General Abacha, powerful and vicious enough to crack down on the oil-producing communities of the Niger Delta whenever the need arose, was best positioned to guarantee this. It is also for broadly the same reasons that the new Labor government in Britain, in spite of its apparent commitment to an "ethical" foreign policy, lobbied Commonwealth countries during the organization's Head of Governments meeting in Edinburgh in September 1997 not to expel Nigeria, as implicitly agreed in 1995 after the murder of Ken Saro-Wiwa, even though the regime had neither improved its appalling human rights record nor moved the country any nearer to democracy.[34]

However, with Western multinational corporations increasingly expanding on a global scale after the collapse of the Soviet Union, and demanding more and unimpeded access to national markets and the world's depleting natural resources, even powerful governments like that of the United States are becoming irrelevant in the face of the "globalization" juggernaut. In June 1998 representatives of the world's twenty-nine wealthiest countries met in

Paris under the auspices of the Organization for Economic Cooperation and Development, hoping to put final touches to an agreement, negotiated since 1995 in great secrecy, to give multinational corporations virtually unlimited powers to run the world as they please. As Renato Ruggerio, Director General of the World Trade Organization (WTO), bluntly put it, "We are writing the constitution of a single global economy."[35] Approval for the agreement, known as the Multilateral Agreement on Investments (MAI), is, however, on hold at the WTO.

If the MAI, already opposed by NGOs worldwide, is enacted, multinational corporations will be able to push national governments aside and invest their capital wherever they want. They will be able to sue "recalcitrant" governments for loss of profits resulting from legislation that ostensibly discriminates against them. International treaties and conventions that protect the world's increasingly threatened ecosystem will be ignored as the big Western corporations that control the world trade in food, oil, and weapons move to gobble up natural resources wherever they can find them. The poor and exploited countries of the South, who were not consulted on the contents of MAI, would be corralled into signing the agreement on the pain of missing out on the so-called foreign direct-investment bonanza. With MAI firmly in place, the environmental human rights of the communities of the Niger Delta would be crushed underfoot, while giant corporations like Shell could be in a position to do as it pleases with their land, unchecked.

The Eye of the Earth

It is this new global threat that environmentalists, democracy activists, community workers, and plain ordinary folk in the oil-producing communities of the Niger Delta are fighting peacefully to defang. The human ecosystem of the Niger Delta is the eye of the earth, tender, fragile, and requiring the utmost care. It has not been given this care by Shell and the succession of military and civilian fortune hunters that pass for government in Nigeria. The World Wide Fund for Nature (WWF) estimates that the direct cost of cleaning up Shell's oil operations in the Niger Delta, not including the company's gas flares, which even now are still poisoning the atmosphere, could reach $6 billion.[36] Ultimately, it is up to the people of the Niger

Delta to protect their most valuable heritage—their environment—from further blight. Shell, still exploring and producing oil in their midst, is a clear and present danger, to their environment, to their lives, and to generations yet unborn.

The danger is very real, and under no circumstances must it be underestimated, negotiated, or trivialized. In Britain, oil contaminated 125 miles of coastline in the South Pembrokeshire area during the Sea Empress oil spill disaster in February 1996. People living within half a mile of the polluted beaches are still suffering from physical and psychological illnesses.[37] How many Sea Empresses have been visited, unchronicled, on the fragile human ecosystem of the Niger Delta? How many poor, hapless villagers have been poisoned to death in these modern-day gas chambers? Yes, let us say it. The people of the Niger Delta, after the long terrible night of slavery, have another holocaust on their hands.

Knowledge is power, and a people sufficiently equipped with this powerful but nonviolent asset will not perish. The people of the Niger Delta, one and all, must become legal experts and of necessity find within the law solutions to the ecological problems facing their threatened ecosystem. They must, like their ancestors, become competent environmentalists, able to master the essential principles of environmental impact assessments and thus take the juggernaut in their midst to task before a single blade of grass is maimed or murdered in the name of oil. They must become political activists, pressing peacefully but fearlessly for self-determination in a new Nigerian federation informed by justice, equity, and negotiated cooperation. Above all, they must, like Ken Saro-Wiwa, become poets and tellers of tales, taking their case against Shell and the Nigerian government before the world.

For other Nigerians, and indeed the world at large, the timeless thoughts of that worthy son of Africa, Frantz Fanon, will suffice: The future will have no pity for those who, possessing the exceptional privilege to speak words of truth to the oppressor, instead take refuge in cynical indifference and cold complicity. Today it is the people of the Niger Delta. Who will it be next? The struggle going on today in the Niger Delta must not be allowed to die. If the Niger Delta dies, Nigeria dies too. The world dies too.

EPILOGUE

"Shell Is Here on Trial"

Excerpts from Ken Saro-Wiwa's pre-conviction statement to the Justice Auta Tribunal:

My Lord, since my arrest on the 21st of May 1994, I have been subjected to physical and mental torture, held incommunicado and denied food for weeks and medical attention for months. My seventy-four-year-old mother has been whipped and arrested, my wife beaten and threatened with detention, the three telephone lines to my office and residence cut, and they remain cut to this day. My office and home have been ransacked on three different occasions, and personal and family property, official files and documents, taken away without documentation.

I have been calumniated in the press and on satellite television before the world by a Rivers State government anxious to prejudice the mind of the public and to convince the public of my guilt even before trial. Only recently, before the United Nations Committee for the Eradication of Racism and Discrimination in Geneva, an official delegation of the Federal Government which included the Special Adviser on Legal Affairs to the Head of State, Professor Yadudu, declared me responsible for the murders which are the subject of his Tribunal, even before the Tribunal has found anything against me or anyone else.

The fact that a case of homicide is being charged before a Tribunal set up Under Decree No. 2 of 1987 speaks for itself. I am aware of the many strictures laid against the decree and this Tribunal by local and international observers. All the same, I have followed the proceeding here with keen and detailed interest, not only because I am charged before this Tribunal, but also because, as a writer, I am a custodian of the conscience of society. I regret that the legal counsel I freely chose, Gani Fawehinmi, the human rights hero and pride of this country, was forced to withdraw. His withdrawal has denied credibility to this trial.

. . . My Lord, we all stand before history. I am a man of peace, of ideas. Appalled by the denigrating poverty of my people, who live on a richly endowed land, distressed by their political marginalization and economic strangulation, angered by the devastation of their land, their ultimate heritage, anxious to preserve their right to life and to a decent living, and determined to usher into this country as a whole a fair and just democratic system which protects every one and every ethnic group and gives us all a valid claim to human civilization, I have devoted all my intellectual and material resources, my very life, to a cause in which I have total belief and from which I cannot be blackmailed or intimidated. I have no doubt at all about the ultimate success of my cause, no matter the trials and tribulations which I and those who believe with me may encounter on our journey. Nor imprisonment nor death can stop our ultimate victory.

I repeat that we all stand before history. I and my colleagues are not the only ones on trial. Shell is here on trial and it is as well that it is represented by counsel said to be holding a watching brief. The company has, indeed, ducked this particular trial, but its day will surely come and the lessons learnt here may prove useful to it, for there is no doubt in my mind that the ecological war the Company has waged in the Delta will be called to question sooner than later and the crimes of that war be duly punished. The crime of the Company's dirty war against the Ogoni people will also be punished.

On trial also is the Nigerian nation, its present rulers and all those who assist them. Any nation which can do to the weak and disadvantaged what the Nigerian nation has done to the Ogoni, loses a claim to independence and to freedom from outside influence. I am not one of those who shy away from protesting injustice and oppression, arguing that they are expected of a military regime. The military do not act alone. They are supported by a gaggle of politicians, lawyers, judges, academics, and businessmen, all of them hiding under the claim that they are only doing their duty, men and women too afraid to wash their pants of their urine.

We all stand on trial, my lord, for by our actions we have denigrated our country and jeopardized the future of our children. As we subscribe to the subnormal and accept double standards, as we lie and cheat openly, as we protect injustice and oppression, empty our classrooms, degrade our hospitals, as we protect injustice and oppression, fill our stomachs with hunger and elect to make ourselves subservient to those who subscribe to higher standards, pursue the truth, and honor justice, freedom, and hard work.

I predict that the scene here will be played and replayed by generations yet unborn. Some have already cast themselves in the role of villains, some are tragic victims, some still have a chance to redeem themselves. The choice is for each individual.

I predict that a denouement of the riddle of the Niger Delta will soon come. The agenda is being set for this trial. Whether the peaceful ways favored will prevail depends on what the oppressor decides, what signals it sends out to the waiting public.

In my innocence of the false charges I face here, in my utter conviction, I call upon the Ogoni people, the peoples of the Niger Delta, and the oppressed minorities of Nigeria to stand up now and fight fearlessly and peacefully for their rights. History is on their side, God is on their side. For the Holy Quran says in Sura 42, verse 41: "All those who fight when oppressed incur no guilt, but Allah shall punish the oppressor." Come the day.

APPENDIX: JUSTICE ON TRIAL

The world as we know it is dying. Our crops, the fishes in our streams, everything is dying. Where is justice, eh?

A seventy-year-old fisherman in Otuegwe village community in the Niger Delta, where a Shell 800,000-barrel oil spill laid waste to vegetation and fish life in 1998

In November 1996 the families of Ken Saro-Wiwa and John Kpuinen, with the support of the New York-based Center for Constitutional Rights, sued Royal Dutch Petroleum Company and Shell Transport and Trading Co. Plc in the U.S. District Court (Southern District of New York). In their depositions, the two families alleged that the executions in November 1995 were carried out with the "knowledge, consent and/or support of Shell." They also said Shell was part of a conspiracy "to violently and ruthlessly suppress any opposition to [the company's] conduct in its exploitation of oil and natural gas resources in Ogoni and in the Niger Delta."[1]

Shell, master of scenario-planning, had anticipated this move. Eight months before the Saro-Wiwa and Kpuinen families went to court, the company had assembled a team of lawyers and environmentalists under the name Environmental Clinic Nigeria/Shell Team and charged it with preparing a detailed memorandum on the outline of environmental problems in its Nigerian operations and potentially applicable law in anticipation of litigation. The team's interim report, handed in on February 26, 1996, was in two parts—Environmental Situation, and Environmental Law and Its Application. After an extensive survey of the relevant international environmental laws, the memorandum also considered applicable Nigerian law and concluded: "The Nigerian Constitution contains a provision espousing a right to protection of the environment. Note that these rights are rarely justiciable."[2]

Thus, Shell's behavior in the Niger Delta have always been dictated by the conviction that its actions against the people of the Niger Delta and their environment cannot be successfully challenged in a Nigerian court of law because the enabling instruments are either absent or, where they exist, can be flouted with impunity.

The Structure of Nigeria's Environmental Laws and Regulations[3]

Nigeria's environmental laws are not to be found in any one volume in the country.[4] They can be ferreted out of several sources, as indeed is the case with the Nigerian legal system.[5] These sources, not necessarily in order of importance, include:

1. English Law. This consists of:
   a. The received English law comprising (i) the common law, (ii) the doctrines of equity, (iii) the statutes of general application in force in England on January 1, 1900, (iv) statutes and subsidiary legislation on specified matters
   b. English law made before October 1, 1960, and extending to Nigeria
2. Nigerian legislation, which consists of (i) ordinances, (ii) acts, (iii) laws, (iv) decrees, (v) edicts
3. Customary law
4. Nigerian case law (judicial precedents)

Obviously, some aspects of English law still form part of Nigerian law, especially in the core area of the Niger Delta, where Shell Petroleum Development Company (SPDC) reigns supreme.[6] The provisions of the Rivers State Edict of 1988 declares:

> An edict to make provision in relation to the application within the state of the common law and the doctrine of equity and to make applicable certain Imperial Acts which are statutes enacted by the parliament of England, the parliament of Great Britain, the parliament of the United Kingdom of Great Britain and Northern Ireland.[7]

Pursuant to these provisions, therefore, Nigeria and indeed the whole territory of the Niger Delta have benefited and are still likely to benefit[8] from the evolving environmental law in Great Britain and Northern Ireland.[9] Nigerian legislation are laws enacted by the various tiers of government within the Nigerian federation.[10] Statutes enacted before October 1, 1954, by the central legislature are called ordinances.[11] As soon as Nigeria became independent from Great Britain, all such ordinances were redesignated Acts.[12] Also, all laws made by the Parliament at the federal level after independence are called acts. On January 15, 1966, a group of army officers violently overthrew the civilian government that took over from the colonial administration. It was not, however, until the next day that a clear position as to who was in charge was ascertained. The new military rulers then began ruling by decrees. Decrees today are the supreme laws of the land.[13] Laws are enactments made by a regional legislature or a state before military rule. Edicts are laws made by the military governor of a state or region during the pendency of military dictatorship.

Customary laws are laws in existence in all the various rural communities of Nigeria. It is now generally agreed that what constitutes a customary rule is that which enjoys widespread acceptability in that community.[14] G. E. Ezejiofor ventures a definition:

> Customary law is a body of customs and traditions which regulate the various kinds of relationships between members of the community in their traditional setting.[15]

They originate from moral rules, shaped by "ancestral beliefs" and now engrafted as "jural postulates"[16] in Nigerian legal jurisprudence. The benefit of customary law to the overall well-being of humankind has been recognized and extolled.[17] As recently as 1990, the Supreme Court declared unambiguously in *Oyewunmi* v. *Ogunesan*:

> Customary law is the organic or living law of the indigenous people of Nigeria, regulating their lives and transactions. It is organic in that it is not static. It is regulatory in that it controls the lives and transactions of the community subject to it. It is said that customs is the mirror of the culture of the people. I would say that customary law goes further and imports justice to the lives of all those subject to it.[18]

Such customary rules, though not written, regulate the use of the land, water, forest resources, fisheries, wildlife, waste disposal, inheritance, marriage, religious beliefs, art, and relationships in society, among others. The current debate in Europe as to whether "animals" should be accorded "rights"[19] just as humankind, is not novel in the Niger Delta. In 1846 a treaty between two indigenous communities of the Niger Delta, namely Bonny and Andoni, declared the need to allow "animal liberty" in the territories of these two neighbors.[20] However, a rule of custom cannot operate as law when it is repugnant to natural justice, equity, and good conscience. As such, the court will declare such laws void because it will not waste its time transforming "a barbarous custom into a milder one."[21] Rules of custom that question the supremacy of the military government in present-day Nigeria are similarly voided.[22]

Nigeria's environmental laws can also be found in judicial precedents laid down by the courts. A precedent is a legal principle on which a judicial decision is based. A combination of all these in addition to the Constitution of Nigeria make up the sources of Nigerian law.[23]

Environmental Outlaw in the Niger Delta

Shell claims that it operates "within the laws of Nigeria," that it does not take political positions, and that it has "never violated any laws" of Nigeria.[24] The company, however, now accepts that "there are problems in the Delta and we are committed to dealing with them."[25] Also, the company has "offered" to clean up Ogoni, one of the several communities where Shell's operations has devastated the ecosystem.[26] We propose to appraise these issues under three broad headings and a fourth subheading. We shall assess environmental laws relating to land, the aqueous environment, and the atmosphere. We shall also consider Nigeria's compensation laws, and specific instances where Shell appears to have violated them.

Violating Laws Protecting Land[27]

Laws protecting land in Nigeria, and by extension the Niger Delta where Shell's activities have the greatest impact, dates back to 1915, with the introduction of the Water Works Act by the colonial administration.[28] Following the 1915 act was the Public Health Act of 1917.[29] Shell came to Nigeria in 1937 and left at the outbreak of the European interethnic war of 1939–1945.[30] The company returned to Nigeria at the end of hostilities in

Europe and discovered oil in "commercial quantities" in the central Delta in 1956.[31]

Between 1945, when the war ended, and 1996 over one hundred laws with relevance to the protection of land were enacted or decreed.[32] Some of the laws are draconian. An example is the Land Use Act, which purports to divest ownership of land from local communities and places such ownership in the hands of the federal government. The government thus became a trustee of land for the people.[33] Other relevant laws are the Environmental Impact Assessment Decree of 1992 and the law that prohibits the importation of plant seed, soils, and containers into Nigeria that could harm the land.[34]

The laws referred to above do not include the over one thousand customary laws regulating the protection of the environment as a whole.[35] Let us consider these customary laws in some detail under three subheadings: forests, wildlife, and soil.

Forests

Forests[36] and forest reserves[37] exist throughout the Niger Delta. National, state, and customary laws exist to regulate the protection, use, preservation, and conservation of these forests.[38] In Rivers and Bayelsa states, the applicable state law is the Forest Law of 1956.[39] The law exists to preserve and control the exploitation of forests and prevent them from unwarranted degradation. Similar provisions are to be found in Edo, Delta, and Ondo state laws, so too in Akwa Ibom, Cross River, Imo, and Abia states.[40]

Customary law regulates the protection of forests in many ways, principal among which are the communal declaration of certain forests and groves as sacred; the delineation of forests as burial grounds for good and evil people (the bad bush practice); the recognition given to and observed in boundary forests between neighboring communities, family heritage forests, forests of common use, and the essential habitat forests. The last deserves some explanation.

A forest habitat that does not fall into any of the other general forest-protection practices may become protected as a result of an event. The protection is not permanent, as other events could occur to erode what led to its protection in the first place. For example, among the Kalabari, Okrika, Nembe, Bonny, Abureni, and Brass people, the Odumu (royal python) is regarded as sacred and is in fact worshiped. Now, should an Odumu find suitable habitation in a part of forest of common use and remain there to reproduce, that part of the forest automatically becomes a "no go area" for members of the community after due consultation with the priest or priestess.[41]

In carrying out operations relating to exploration for oil, SPDC employs seismographic companies[42] that cut seismic lines through tropical rain forest, swamp forests, mangroves, and farmland. The lines are sometimes as wide as fifteen feet and run into hundreds of miles in length. "No tree no matter how big was allowed to stand in Shell's way," remarked Chief Irene Amangala, who lives in Oloibiri.[43] In Okoroba the phrase "as straight as Shell's road" has gained currency and aptly describes the company's seismic lines and its overarching dominance of the people's daily life.[44] In the process of constructing pipelines to transport crude from flow stations[45] to the tank farms located in Forcados, Bonny, and Brass, and in constructing pipelines from gas fields to gas plants as in Utorugu,[46] Shell as a matter of routine cuts down thousands of acres of rain forest, mangrove forests, and forests in the barrier islands.[47] These lines, grotesque and ugly, are visible even from the air. They crisscross the Niger Delta landscape like the chaotic markings of a demented cartographer.

The result of this assault is that previously inaccessible areas of dense forest were opened up to illegal hunters and timber loggers, who have invaded the area in the past thirty years and carted away invaluable resources.[48] The proposed Taylor Creek forest reserve has almost lost its viability to ceaseless seismographic and other related oil industry activities directed and controlled by Shell. J. P. Van Dessel, SPDC's former head of environmental studies, told ERA activists in October 1995 that during the visit of the Prince of Wales to the area in 1991, Shell had pledged to establish a forest reserve in Taylor Creek, an ecologically interesting ecosystem with a thriving population of monkeys, elephants, chimpanzees, crocodiles, and pigmy hippopotami. Nothing came of the venture, however. The company has admitted that it is a historic polluter of land in the Niger Delta, particularly freshwater swamp forests, mangrove forests, high forests, and forests in the barrier islands. As recently as 1994, Shell told a World Bank team that fifty-five incidents of oil spills, spewing 515 barrels of oil into the environment, occurred in the Western Division of its operations. In the Eastern Division, 203 incidents resulting in 18,527 barrels of crude oil being spilled in forest areas were recorded.[49]

This is almost double the quantity spilled by the other oil companies operating in the area put together.[50]

Have these activities by Shell violated any of the laws protecting forests or forest reserves in the Niger Delta? Section 30 (1) of the 1979 Constitution guarantees the right to life of all Nigerians, including those living in the Niger Delta. The state or any person is not to do anything that impairs or takes away the lives of the people of the Niger Delta. The inhabitants are in the main agrarian, and they still gather food.[51] They depend to a large extent on what the various forests hold in store for them.[52] Their survival is closely linked with the survival of the forests. To destroy the forests, as Shell has consistently done through seismic and other oil industry activities,[53] is to deprive the people of their means of livelihood. Shell has set parts of the Niger Delta over which it superintends on the path of death.[54] Other laws that Shell's activities have probably violated include provisions in the criminal code governing the prevention of nuisance,[55] deposition or discharge of harmful waste on land,[56] and Federal Environmental Protection Agency (FEPA) Decree 1988.[57] The FEPA decree spells out liability and penalties for spillers of hazardous substances whether on water or land. Such spillers are compelled by the decree to bear the cost of removal, replacement of natural resources damaged or destroyed by the discharge, and report same to the agency or other related agencies.[58] No record of Shell reporting the spillage at Iko exists in FEPA records.[59] Also, there is no area in the Delta where Shell has successfully restored or replaced a natural ecosystem.[60] The company, for forty years, failed to support any conservation program of note in the Niger Delta, preferring to support the high-profile WWF-funded Cross River National Park located near the Cameroon border.[61]

Shell's destruction of the Niger Delta environment has been recorded by local and international observers alike, including environmental groups and sundry agencies.[62] What has not been reported is the multinational's disrespect and flagrant violations of local customs and law. These laws were developed from moral rules that still exist in the communities of the Niger Delta. Environmentalists and public commentators who ask Shell to rethink and overhaul its policies so they are in line with the way of life of these communities have often been chastised by the company's executives in Lagos, London, and The Hague. The former chairman of the Shell Group,

C. A. J. Herkstroter, dismissed these views as emanating from "moral imperialists and moral relativists," referring to voices critical of his company's practices in Europe and Nigeria respectively.

In refusing to respect local laws and customs relating to the environment, Shell violates Regulation 45 of the Petroleum (Drilling and Production) Regulation. The law enjoins oil companies to ensure that their activities do not "hinder" the development of the communities.[63] Specifically, Regulation 50 requires oil companies to respect "communal areas or objects declared sacred by the community or government as well as tree and mangroves venerated under their custom. These should not be destroyed by the operations of the oil company."[64]

Shell is in a joint venture arrangement with the government of Nigeria in the exploration, exploitation, production, and marketing of crude oil and petroleum products.[65] In the joint venture, the Nigerian government is only interested in how much money is due it.[66] Nigeria, as a member of the international community, is a signatory to several international legal instruments, and is expected to observe and respect such obligations. Shell, by its operations, has contributed to Nigeria's violation of several of these laws, particularly those relevant to biodiversity conservation.[67] The conventions violated include the International Convention for the Prevention of Pollution of the Sea by Oil, 1954; Convention on Fishing and Conservation of Nature and Natural Resources of the High Seas, 1966; African Convention on Conservation of Nature and Natural Resources, 1968; the African Charter on Human and Peoples' Rights; Convention for Cooperation in the Protection and Development of Marine and Coastal Environment of the West and Central African Region, 1984; Convention on the Conservation of Migratory Species of Wild Animals (Bonn Convention), 1979; and the Base Convention on the Control of Transboundary Movements of Hazardous Wastes and Their Disposal, 1989; among others.

Eboe Hutchful said of oil industry pollution: "Environmental pollution from the oil industry has had far-reaching effects on the organization of peasant life and production. In addition to the effects of spills on mangroves already noted, spills of crude, dumping of by-products for exploration, exploitation and refining operations (often in freshwater environments), and overflowing of oil wastes in burrow pits during heavy rains has had deleterious effects on bodies of surface water used for drinking, fishing, and household and industrial purposes . . . Spills, disposal of industry by-products, and flaring of gas also have had widespread repercussions on the availability and production of farming land."[68]

The NNPC, Shell's venture partner in the Niger Delta, agrees that oil companies are wreaking havoc on the environment.[69] By the admission of these two companies, crude oil emanating from frequent spillage does enter the sea through estuarine currents and then are driven westward and eastward by the Guinea currents. Shell also dumps household waste (poisonous food not fit for human consumption) and hazardous waste (oil from generators and drilling mud) directly into the sea, thus violating the London Dumping Convention and the Base Convention on Transboundary Movements of Hazardous Waste and Their Disposal, 1989.[70] One international legal instrument that has suffered the greatest violation is the African Charter on Human and Peoples' Rights. Article 24 guarantees all peoples of Africa, including the peoples of the Niger Delta, the "right to a satisfactory environment favorable to their development." The development of the Niger Delta people includes the protection of their land, culture, and customs.[71] Gas flares have banished nights as infernal blazes light up the Delta skyline in Shell's areas of operations, killing moths, butterflies, grasshoppers, and other valuable insects whose usefulness to ecological stability is beyond question. Not a single one of Shell's gas-flaring nozzles is protected.[72] Oil pollution is the biggest contributor to the impoverishment of people in Shell's oil fields.

Wildlife Protection, Animal Rights

The concept of animal libertarianism has been practiced among the communities of the Niger Delta since antiquity. The concept received formal legislative stamp in 1846, when the first recorded treaty between two neighboring communities (Bonny and Andoni) was signed. The effect of the treaty has benefited the world today, as the largest population of elephants in the Niger Delta is still found in the Bonny-Andoni area. Because of the importance of animals to the lives of the people, we shall reproduce two relevant articles of the treaty, the sixth and twelfth, respectively:

> They, the Andony men, also promise not to destroy the Guano [iguana] but allow animal liberty the same in Bonny.
>
> Should the Andony men kill any elephants, they are to present the teeth thereof to King Pepple; and should the Andony men at any time be short of musket or powder, King Pepple will supply them.[73]

Respect for animals began as early as humanity began settled life in the Niger Delta more than five thousand years ago.[74] Humankind have lived and developed a harmonious relationship with their environment, even bordering on veneration. This is attested to by the Christian historian Tasie:

> The region lies in the low equatorial forest plain of Nigeria, with the coastal area shaded by stately and changeless evergreen mangrove which disappears inland into a zone of tropical forestland with some luxuriant foliage of various huge trees and almost impenetrable thick undergrowth. The plant life include the cotton tree, African oak and Iroko, growing to such enormous size that in certain parts of the Delta they are set apart as objects of worship or veneration. One finds an enormous variety of birds, animals, and reptiles. Among the last named, the python Sebae or African python is widely venerated.[75]

Sir Alan Burns, the British historian, chronicled events in Brass and Bonny thus:

> At Bonny and Brass, for instance, the monitor lizard and the python respectively were regarded as sacred and were allowed to crawl throughout the towns, no one being allowed to kill or in any way interfere with them; so real was this animal worship till about 1878 that British subjects were actually fined by the Consul for molesting the sacred reptiles.[76]

The dynamic nature of customary law has ensured that obnoxious laws are weeded out with time.[77] In the Niger Delta, respect for animals has survived this dynamism as far as written records in the area can confirm.[78] The people do not hunt animals for sport; rather, they are categorized in accordance with value—religious, ecological, social, and economic.[79] Of these, only the economic imperative allows members of a community to kill an animal, either as food or as a nuisance and threat to societal harmony that has to be removed.[80] Thus, grass cutters, rabbits, leopards, bush cows, and a few other animals have always been killed for food. Other animals such as crocodiles, iguana, chimpanzees, some species of monkeys, several types of birds (the grey parrot, eagles, the owl, the fishing owl, bats, herons, doves, blue-breasted kingfisher, etc.) and some snakes are guarded by customary laws and protected from harm. In earlier times the killing of any animal on the prohibited list, either willfully or accidentally, attracted the death penalty:

> In the year 1787 two of the seamen of a Liverpool ship trading at Bonny, being ashore watering, had the misfortune to kill a guana as they were rolling a cask to the beach . . . The offenders being carried before the King or chiefman of the place were adjudged to die. However, the severity of justice being softed by a bribe from the captain, the sentence was at length changed to the following, that they should pay a fine of 700 bars (about £175) and remain in the country as slaves to the King until the money should be raised.[81]

This narration contains an error, though. The "bribe" was not a bribe as such but money required to buy materials to carry out the necessary sacrifice to cleanse the town and thus ward off the calamities that otherwise would befall it following the killing of a sacred creature.

We have chosen to go to this length to demonstrate how seriously the various communities of the Niger Delta take animal protection, a practice that finds eloquent expression in their norms, values, and customary laws. Flowing naturally from this, subsequent enactments in this area of law at the national level have to a large extent found favor in the eyes of the communities.[82] These laws include: the Wild Animals Preservation Law, 1916, the Mosquitoes Destruction Law,[83] and the Disease of Animals Law, 1917.[84] The last two accord with the nuisance value appellation already in use by several of the Niger Delta communities.

So has Shell extended its tentacles into these sacred forests and groves and violated the "liberty" of animals as well? Very much so—by allowing harmful wastes to be introduced into the habitat[85] of these animals through seismic and pipeline-laying activities, and refusing to strictly regulate the work of contractors who go on the rampage[86] destroying economic trees and food crops on which animals depend. To preserve an animal is to allow it to live until it dies a natural death. The Wildlife Act supports this, except when there are authorized killings by competent authorities in cases of disease or overpopulation. For communities whose very way of life is closely interwoven with some of these animals, the attendant misery of living with a lawless juggernaut like Shell these past forty years would be simply difficult to assess.

Maiming the Soil

Soil in this context is the fertile topsoil (two to eight inches) that crops planted by the local communities thrive on. Agricultural production in the Delta takes place mostly on levees during the nonflooding season, on mudflats or the banks of rivers (rich in humus), and on naturally drained land. Customary, national, and international laws protect these practices and the soil so that humanity in this part of the world can survive.[87]

Shell is known to have carried out activities in the Peremabiri area of the Niger Delta that have caused massive erosion of streams, rivers, and land. Houses were also ruined. In Nembe, eleven incidents of oil spillage occurred between December 10, 1994, and February 10, 1995, that affected the town[88] and the surrounding environment as far as the farming communities of Agrisaba, Egenelogu, Ikensi, Odioma, Elemuama, Idema, and Obiata. Botem Tai, Bomu, Korokoro, Kpite, K-Dere, and the riverbank communities, principally Kaa, have had had their soils ravaged by the destructive effects of oil pollution.[89] E. J. Fekunmo, a senior lecturer in law at the University of Science and Technology in Port Harcourt, described this kind of pollution as "deliberate infliction of injury to persons and property" and called for changes to the traditional rules of liability, which have provisions that favor Shell and the other oil companies and cheat the local communities. However, given the bitter experience of the communities these past forty years, it is most unlikely that Shell will mend its ways even when these rules are further strenghtened.[90]

Shell and the Violation of the Aqueous Environment

Aqueous environmental laws are legal rules that protect rivers, lakes, wells, ponds, estuaries, fish dams, fishing grounds, and the general biodiversity of aquabased life from the polluting activities of Shell and other such agents. These laws include: the Oil in Navigable Waters Decree, the Sea Fisheries Act, the Criminal Code, the Harmful Waste (Special Criminal Provisions, etc.) Decree, the River Basins Development Authority Act, the Petroleum Act of 1969, the Petroleum (Drilling and Production) Regulation Decree, the Federal Environmental Protection Agency Decree, the Works Act of 1915, the Public Health Act of 1917, the Territorial Waters (Amendment) Act 102, and the Explosives Act of 1967, all federal laws. There is also the Explosive Law 104 (applicable in Bayelsa and Rivers states) and another federal law—the Oil Pipelines Act of October 4, 1956.

Local customary laws with respect to the protection of the aqueous environment exist in several communities in many ways: sacred rivers, ponds, and lakes;[91] rivers and lakes of seasonal usage;[92] and laws with respect to individual species in a particular or general aqueous environment.[93] There are criminal and civil liabilities recognized under customary laws against the destruction of some or all of these prohibited species. Infraction could result in fines, expensive ablution and atonement rites, and ostracism from the age group or the community as a whole.

One clear example of Shell's violation of the aqueous environment with impunity is the case of ***Allar Irou*** v. ***Shell-BP***, in which the judge refused to grant an injunction in favor of the plaintiff, who had suffered enormous losses in the form of polluted land, fishponds, and dead fishes, arising from the defendant's activities in the course of the company's exploitation of mineral crude oil. M. A. Ajomo, director of the Institute of Advanced Legal Studies, in Lagos, commented on this case: "In the oil sector where environmental degradation is mostly prevalent, the all-pervading influence of the oil companies and the paternalistic attitude of the judges toward them in matters relating to environmental hazards created by these companies have made the enforcement of environmental laws ineffective."[94]

The acts of Shell in polluting Allar Irou's land is contrary to Section 11 (5)C of the Oil Pipelines Act, Cap 145, of 1958. The statute creates strict liability in this kind of pollution. In paragraph 36 of Schedule 1 of the Petroleum Act of 1969, the law says:

> A holder of an oil exploration license, oil prospecting license or oil mining lease shall in addition to any liability for compensation to which he may be subject under any provisions of this Act be liable to fair and adequate compensation for the disturbance of surface or other rights to any person who owns leased land.

The overwhelming majority of the communities of the Niger Delta do not have the financial resources to drag Shell before the law courts. They simply suffer in silence and hope for justice on Judgment Day.[95]

The most serious threat to the aqueous environment is the threat posed to aqueous life by the use of explosives. Explosives and mines are known to have maimed and killed humans and destabilized the ecosystem in southern Africa. In the Niger Delta, marine life is under severe threat because of the activities of Shell and the other oil companies operating in the area. Dynamite is used by the companies during seismographic activities as they explore for oil deposits. A seismic shooter employed by a German seismographic company in Port Harcourt described the process:

> In very simple terms, we lay the seismic cables along the route and connect the necessary positive and negative points of impact. Then we retreat to a maximum allowable safety regulation distance, and through a remote detonator we set the system to work. A massive explosion which shakes the whole area for miles then follows. Trees do get uprooted and thrown to the ground. Sometimes huge chunks of earth are lifted and thrown into the rivers, or a land area can be cut off from other land and if it is close to a river an island is created. The explosion scares birds, kills animals, and sometimes destroys buildings.[96]

What the seismographer did not mention was the effect on fish generally, especially during the breeding season. It is also significant that some communities in the Niger Delta that are now experiencing earth tremors, in an otherwise earthquake-free area, have suffered continuous blasting of the earth crust during oil-related activities.[97] Even more worrisome is the illegal sale of dynamite to inhabitants of local communities, who now use the device to kill fish. According to a former employee of the British Seismographic Services Ltd. now taken over by the German firm Geko Prakla Schlumberger:

> When we were in SSL, we used dynamite to kill fish. We simply connect the device to any lake or river and let go. All fishes within the area of the blast die. We select the big ones and leave the small ones. Sometimes we sell some of the dynamite to make money or simply give them away to our friends in the area where our camps or houseboats are located. We often do this when we have industrial disputes or to please the youth of our host communities.[98]

The Oil in Navigable Waters Act prohibits discharge of crude oil and heavy diesel in the waters of Nigeria.[99] The former chief executive of SPDC, Brian Anderson, admitted in London that "we do not have a waste treatment plant" in the Niger Delta.[100] According to Anderson, his company had just ordered one from Canada. So where has Shell been dumping its heavy disused diesel oil waste from giant generators in flow stations; residential houseboats anchored in the sea, rivers, and estuaries; waste oil from speedboats owned and operated by contractors but supervised by the multinational? The truth is that Shell has turned the aqueous environment of the Delta into one huge dump site for the disposal of waste.[101]

Methods of spill control and management do not accord with that recommended by Regulation 7 of the Minerals Oil Safety Regulations of "good oil field practice" as in current use by the Institute of Petroleum Safety Codes, the American Petroleum Institute Codes, and the American Society of Mechanical Engineers. The regulation requires oil companies to adopt such practice.[102] In 1994 a team of environmentalists visited Shell Camp in Port Harcourt and saw an oil dump site where vehicles receive petroleum products and staff collect other oil-based products for their work. A lot of oil is routinely spilled here, and the dump is very close to a drinking water bore hole.[103]

Atmosphere and Noise Pollution

Atmospheric pollution by Shell constitutes the biggest source of worry for the international community. It is visible and easily documented, and Shell has admitted to polluting the biosphere through continuous gas-flaring.[104]

The company's action is therefore a deliberate infraction of enactments, some of which are criminal provisions.[105] The offense of public nuisance under Section 234 of the Criminal Code does not allow Shell to cause "any inconvenience or damage to the public in the exercise of rights common to all members of the public."[106] Shell's right to explore and produce hydrocarbon (a right now seriously challenged by all communities of the Niger Delta) does not preclude the rights recognized to be enjoyed by persons, animals, and other living and nonliving things. The 1985 Rivers State Edict prohibits noise from vehicles, music shops, etc. The law is also applicable in Bayelsa State. Noise from vehicles "constitutes the bulk of the noise produced by vehicular traffic in the metropolis of Port Harcourt, Warri, and Yenogoa," according to Kemedi Demeiri, an environmental activist. Oil companies and their workers control more than a quarter of the vehicles in the Niger Delta. Add the noise to the huge decibels from generators, seismographic activities, music from camps and houseboats, and you will get a fair picture of how Shell—which generates some 50 percent of these activities—has converted a hitherto serene human ecosystem into one massive bedlam. Section 34 of the FEPA decree prohibits the discharge of hazardous substances into the air. Section 19 controls noise, and Section 21 spells out liability for the infraction. Gas flaring at the Utapete flow station at Iko in Akwa Ibom State does not satisfy the requirements of the relevant regulations. Constantly, bund walls break and the nozzles corrode, creating directionless flare points. Incidentally, the village of Iko, where thousands of human beings live, is less than a quarter mile away. The walls of the local Qua Iboe church, built at Iko in the nineteenth century, have virtually disintegrated after years of relentless bombardment of noxious gases from Shell's Utapete flow station.

Customary laws protect the environment from noise and air pollution. In local architecture, chimneys are not built "in a manner as to cause inconvenience to human habitation or endanger objects of veneration."[107] The flawed engineering of Shell's flow station in Iko is made worse by its location as the flaring funnels are "aided to devastating effect on the inhabitants by the inward driving wind of the Atlantic ocean. The wind directs these hazardous substances on people, animals, and on building materials."[108] This violates the right to life,[109] the right to a satisfactory environment,[110] the relevant portions of the FEPA decree, the Petroleum Act, and the regulations enacted pursuant to that law. The acts of Shell also violate the Mosquito Destruction Law of 1945, because as Isaac Osuaka, an environmentalist who heads an NGO in Port Harcourt, explained, "the flares keep the whole community permanently in daylight, and the only places where they can hide are people's homes where inhabitants try to 'create' night. Mosquitoes love darkness."[111]

Violation of Compensation Laws

Shell regularly tells foreign journalists and environmentalists that it pays adequate compensation and that it is usually above the government-approved rate.[112] Payment of compensation does not satisfy government regulations even if it is above rates set by government. The law enjoins oil companies to pay "fair and adequate compensation" to all those who may be affected by activities of oil operation owners and occupiers of land.[113]

Chief Egi-Ikata of Nembe told journalists in 1994 that what Shell pays the communities as compensation is a pittance and not enough to replace what was destroyed. *The Guardian* of Lagos subsequently reported:

> After Shell have been informed of an oil spillage, its officials would inspect the surface area affected before repairs are made. They will then carry away the damaged nets for inspection, to determine the extent of damage and how much compensation to pay . . . When the compensation finally comes, Shell pays N1,000 for a fishing net which costs N5,000 in the market. Three months ago this was increased to N2,500 per net.[114]

Shell's attitude to compensation claims is the same all through its area of operations.[115] The Court of Appeal found Shell wanting in its compensation practice and stated that the Petroleum Act requires Shell to pay adequate compensation to Farah and all those who have suffered from the "decertification" of agrarian land by Shell's oil pollution. The case emphasized that what is paid must be in accordance with current market value. If the decision of the Farah case is read along with the FEPA decree, Shell ought to restore all polluted land in all its areas of operation to its natural state. In several cases, Shell has tried to evade compensation by hiding under technical rules. In *San Ikpede* v. *Shell-BP* it was successfully argued that the matter of pollution did not come under the rule of strict liability as demonstrated in the common law principle of *Rylands* v. *Fletcher.*[116] Justice Ovie Whiskey (who later rose to the Supreme Court) said: "To lay crude oil carrying pipes through swamp forest land is a non-natural use of land" and "It is common knowledge that crude oil causes havoc to fishes and crops if allowed to escape from the pipeline in which it is being carried."[117]

The judge found Shell guilty of violating section 11(5)c of the Oil Pipelines Act, Cap 145 of 1958, on the basis of strict liability and ordered the company to pay adequate compensation. But Chief A. S. Amos and others who sued Shell-BP a few months later were not lucky to have a judge brave and principled enough to resist the multinational's influence. The former were denied compensation when the blocking of the Kolo Creek by Shell caused them great inconvenience. Shell-BP argued that its blocking of the creek was a public nuisance, and that in any case compensation had been paid to the Imiringi people, the fact that the defendants were members of the public who use the waterways notwithstanding. The courts agreed with the multinational even though it was brought to their notice of a breach of the Petroleum (Drilling and Production) Regulation, 1969.[118] It would have also been grounded on trespass, since it is action without proof of damage.[119]

What Can the Legal World Do?

It is clear from the foregoing that environmental justice cannot flow from the Nigerian state. As Godwin Uyi Ojo, an environmentalist, pointed out, "Every government in Nigeria has been in business with Shell, and Shell acts as the unofficial senior partner. As far as petroleum issues are concerned, Shell is widely seen as dictating the pace of legislation and influencing its enforcement to its own advantage."[120]

On January 31, 1996, Shell executives who were summoned before the Irish Parliament conducted themselves as if they were Nigerian government ministers, prompting one of the Irish lawmakers to ask pointedly whether they were indeed Shell officials or representatives of General Sani Abacha.[121] A government effectively under Shell's thumb cannot do anything to rock the boat. Nor is it in its objective interest to do so. Clearly, help is needed from the legal community, both within Nigeria and internationally, particularly with reference to the company's disregard for rules and regulations. To the extent that Shell's misdeeds in the Niger Delta go unpunished, to that same extent is justice on trial in Nigeria.

NOTES AND REFERENCES

Introduction

1. Barry Hillenbrand, "Seeing the Light in Nigeria," *Time*, November 25, 1996.

2. Dofie Ola, "Ake: the Scholar-Activist." *Liberty*, Jan.-April 1997. Professor Claude, who was Director of the Port Harcourt-based think tank Centre for Advanced Social Science (CASS), died in a plane crash in southwestern Nigeria on November 7, 1996.

Chapter One: A PEOPLE AND THEIR ENVIRONMENT

1. Kenneth O. Dike, *Trade and Politics in the Niger Delta 1830-1885. An Introduction to the Economic and Political History of Nigeria* (Oxford University Press, 1956), 21.

2. Michael Crowther, *The Story of Nigeria* (London: Faber and Faber, 1978), 59.

3. Ibid., 60.

4. Hugh Thomas, *The Slave Trade. The History of the Atlantic Slave Trade 1440-1870* (New York: Simon and Schuster, 1997), 21-22.

5. Dike, *Trade and Politics*, 101.

6. Crowther, *The Story of Nigeria*, 123.

7. Dike, *Trade and Politics*, 112.

8. Kenneth O. Dike, "John Beecroft 1790-1854: Her Britannic Majesty's Consul to the Bights of Benin and Biafra 1849-54," in Crowther, *The Story of Nigeria*, 314.

9. See Dike, *Trade and Politics*, Chapter Seven, for details of Beecroft's career.

10. J. F. Ade Ajayi, "The British Occupation of Lagos 1851-1861," *Nigeria Magazine*, No. 69, August 1961, 96-105.

11. See Chapter Seven of Dike, *Trade and Politics*.

12. Foreign Office. F.O. 84/1117, No. 112, Class B, F.O., Russel to Hutchinson, September 4, 1860.

13. Foreign Office. F.O. 2/34, Minutes on Captain Washington's report, Palmerstone, April 22, 1860.

14. Dike, *Trade and Politics*, 198.

15. Foreign Office. F.O. 84/1630, Opobo Town, Ja ja to Lord Granville, April 3, 1882.

16. Foreign Office. F.O. 84/1630, F.O. 84/1617, F.O. 84/1634.

17. Obaro Ikime, *Niger Delta Rivalry* (London: Longmans, 1969), 69.

18. Dike, *Trade and Politics*, 213.

19. Foreign Office. F.O. 84/1617, No. 16 and its enclosures, Hewett to Granville, November 8, 1882.

20. Dike, *Trade and Politics*, 210.

21. Dike, ibid., 209-215.

22. Alan C. Burns, *History of Nigeria* (London: Allen and Unwin, 7th edition, 1956), 157-158.

23. J. E. Flint, *Sir George Goldie and the Making of Nigeria* (London: Macmillan, 1960), 187-215.

24. Foreign Office. *Annual Report for Northern Nigeria*, 1902, Appendix 111; 105.

25. Crowther, *The Story of Nigeria*, 191.

26. Obafemi Awolowo, *Thoughts on Nigerian Constitution* (Ibadan University Press, 2nd edition, 1966), 14.

27. Frederick Lugard, *Reports on Amalgamation*, 14-15, cited in Crowther, *Story*, 200.

28. *Proposals for the Revision of the Constitution of Nigeria* (London, 1945), 6. Cited in Okechurkewu Nebolisa, *Nigeria and the Crisis of Constitutionalism* (Enugu, Nigeria: Stone Press), 78.

29. Kenneth O. Dike, *One Hundred Years of British Rule in Nigeria 1851-1951* (Lagos, 1957), 43.

30. Crowther, *Story*, 197.

31. Obafemi Awolowo, *Thoughts*, 26.

32. See Alexander Madiebo, *The Nigerian Revolution and the Biafra War* (Enugu: Fourth Dimension Publishers, 1980), for a full account of the civil war.

Chapter Two: SOLDIERS, GANGSTERS, AND OIL

1. Chinua Achebe, *Anthills of the Savannah* (Oxford: Heinemann, 1987), 141. Achebe, author of such classic novels as *Things Fall Apart* and *Arrow of God*, among others, is Africa's foremost writer.

2. Billy Dudley, *An Introduction to Nigerian Government and Politics* (Macmillan Nigeria, 1982), 80. It is, however, significant that Nzeogwu told journalists that the intention was to install Chief Obafemi Awolowo, a non-Igbo politician, as Executive President. See Robin Luckham, *The Nigerian Military. A Sociological Analysis of Authority and Revolt 1960-67* (Cambridge University Press, 1971), 42 (footnote 3).

3. Sarah Ahmad Khan, *Nigeria: The Political Economy of Oil* (Oxford University Press, 1994), 10.

4. Wole Soyinka, *Opera Wonyosi* (Ibadan: Collected Plays, 1983), 24.

5. Khan, *Nigeria*, 9.

6. See Michael Leapman, "British Interests, Nigerian Tragedy," *Independent on Sunday*, January 4, 1998.

7. Khan, *Nigeria*, 16.

8. Khan, ibid.

9. Michael Crowther, *The Story of Nigeria* (London: Longmans, 1978), 278.

10. Human Rights Watch/Africa, *Nigeria; The Ogoni Crisis: A Case Study of Military Repression in South Eastern Nigeria* (London, July 1995), 7.

11. Crowther, 242–44.

12. Greenpeace Nederland, *The Niger Delta: A Disrupted Ecology. The Role of Shell and Other Oil Companies*, October 1996, 33.

13. Godwin Namane Loolo, *A History of Ogoni* (Self-published, Port Harcourt, 1981), 46. See also "Movement for the Survival of the Ogoni People (MOSOP)," *Ogoni Bill of Rights* (Port Harcourt: Saros Publishers, 1991), back page.

14. Olusegun Obasanjo, "Broadcast to the Nation on the Report of the Land Use Panel," Lagos, Federal Ministry of Information, 1978, in William D. Graf, *The Nigerian State; Political Economy, State Class and Political System in the Post Colonial Era* (London: James Currey, 1988), 194.

15. William D. Graf, *The Nigerian State; Political Economy, State Class and Political System in the Post Colonial Era* (London: James Currey, 1988) 67–74.

16. Ibid., 141.

17. See Graf, *Nigerian State*, 240, for details of the role of the International Monetary Fund, the World Bank, and Western transnational banks in imposing the Structural Adjustment Program on Nigeria.

18. Chinua Achebe on Nigeria's political economy in his novel *Anthills of the Savannah* (Oxford: Heinemann, 1987), 141.

19. Graf, *Nigerian State*, 9.

20. Dudley, *Nigerian Government*, 113.

21. Graf, *Nigerian State*, 218.

22. Khan, *Nigeria*, 199.

23. Terisa Turner, "Multinational Corporations and the Instability of the Nigerian State," *Review of African Political Economy*, Vol. 5 (1976).

24. See Chapter Four of Dudley, *Nigerian Government*.

25. Claude Ake, *Revolutionary Pressures in Africa* (London: Zed Press, 1978), 49.

26. Gavin Williams, ed., *Nigeria: Economy and Society* (London: Rex Collins, 1976), 13.

27. Graf, *Nigerian State*, 157.

28. Tom Forrest, *Politics and Economic Development in Nigeria* (Oxford: Westview Press, 1993), 171.

29. See *West Africa*, March 26, 1984, 691.

30. See *Newswatch*, January 13, 1986, for details of the Johnson Mathey Bank scandal.

31. See *West Africa*, September 23, 1985; 1988.

32. See *Observer*, Benin City, October 7, 1992.

33. Ken Saro-Wiwa, personal communication, June 1993.

34. Greenpeace Nederland, *The Niger Delta*, 34.

35. Ken Saro-Wiwa, speech delivered on the occasion of a Nigerian government ministerial visit to Ogoni in January 1993.

36. Human Rights Watch/Africa, *Ogoni Crisis*, 39.

37. Mike Akpan, "Hope Betrayed," *Newswatch*, January 27, 1997.

38. Ibid.

39. Michael Birnbaum, *Nigeria: Fundamental Rights Denied. Report of the Trial of Ken Saro-Wiwa and Others* (London: Article 19, London, 1995). See the appendix of this report for details of the two Ogoni prosecution witnesses' allegations against Shell and Nigerian government officials.

40. Rivers State Internal Security Task Force, Internal Memorandum, Government House, Port Harcourt, May 21, 1994.

41. Kayode Fayemi, *The Oil Weapon: Sanctions and the Nigerian Military Regime. A Report on the Need for Full and Comprehensive Sanctions*. London, December 1995.

42. Quoted in Fayemi, *Oil Weapon*, 6.

43. Khan, *Nigeria*, 32.

44. See *The News* cover story of September 12, 1993, for details of the corruption and mismanagement during the Babangida years.

45. Julius Ihonvbere, "Are Things Falling Apart? The Military and the Crisis of Democratisation in Nigeria," *The Journal of Modern African Studies*, Vol. 34 No. 2 (Cambridge, 1996).

46. *Source*, Lagos, October 17, 1997.

47. These allegations were made by the Association of Concerned Professionals, a Nigerian pressure group, and subsequently published in several newspapers in June 1997.

48. Kayode Fayemi, *Oil Weapon*.

49. *Nigeria Now*, Vol. 7 No. 3, June–December 1997.

50. John Okafor, "Another June 12 Looms," *Tell*, March 8, 1999.

51. See Human Rights Watch/Africa, "Nigeria. Permanent Transition: Current Violations of Human Rights in Nigeria," September 1996, for details of the death of Kudirat Abiola and other democracy activists. Nigeria's influential political weekly, *Tell*, also reported that the assassination of Kudirat Abiola was the handiwork of the dreaded K-Squad, commanded by Major Hamza Al-Mustapha, General Abacha's security chief. See Osa Director, "Abacha's Killing Machine," *Tell*, May 3, 1999.

52. Obed Awoweye, "The Abacha Loot. Winking at the Augean Stable," *Tell*, November 23, 1998.

53. Ibid.

54. Dele Agekameh, "They Died for Our Freedom," *Tell*, March 15, 1999.

55. See Human Rights Watch/Africa, "Nigeria. Crackdown in the Niger Delta," May 1999, Vol. 11 No. 2A, 22–23.

56. "Nigeria: Muddy Waters," *Africa Confidential*, April 12, 1999, Vol. 40 No. 7.

57. "Nigeria: The $2.7 Billion Hole in the Bank," *Africa Confidential*, April 16, 1999, Vol. 40 No. 8.

58. Chris McGreal, "Rooting Out Looters," *Guardian*, London, June 15, 1999.

59. Ibid.

60. Felix Tuodolor, President of Ijaw Youth Council (IYC), in an interview with Ike Okonta, June 15, 1999. The Ijaw are the largest ethnic nationality in the Niger Delta.

Chapter Three: COLOSSUS ON THE NIGER

1. Greenpeace International, *Shell-Shocked. The Environmental and Social Costs of Living with Shell in Nigeria* (Researched and written by Andrew Rowell), July 1994, 9.

2. Barry Hillenbrand, "Seeing the Light," 69.

3. *Financial Times*, July 10, 1985.

4. Roger Moody, *The Gulliver File. Mines, People and Land: a Global Battleground* (London: Minewatch, 1992), 702.

5. Quoted in Andrew Rowell, "Unloveable Shell, the Goddess of Oil," *Guardian*, November 15, 1997.

6. See Moody, *Gulliver File*, 703, for details of Shell companies and subsidiaries worldwide.

7. Ibid.

8. See *Financial Times* of March 16, 1985, for the full story of how Shell planned to drill for oil within sight of the Harkam meadows, immortalized by a famous John Constable painting.

9. Moody, ibid., 708.

10. Ibid.

11. *SAICC Newsletter*, Vol. 2 No. 1 (Berkeley, Autumn 1985).

12. Cimra Documentation, London, November 1978, in Moody, *Gulliver File*, 707.

13. Interview by Dutch researcher, quoted in Moody, *Gulliver File*, 707.

14. Moody, ibid.

15. *Taraxacum* (IYF), Vol. 3 No. 2 (Denmark, 1984).

16. Project Underground, *Human Rights and Environmental Operations Information on the Royal Dutch/Shell Group of Companies 1996-1997* (Berkeley, April 1997), 15.

17. Ibid.

18. Pratap Chatterjee, "Marching to a New Mantra," *Guardian*, May 14, 1997.

19. Sue Brandford and Oriel Glock, *The Last Frontier: Fighting Over Land in the Amazon* (London: Zed Books, 1985), quoted in Roger Moody, *Gulliver File*, 706-707.

20. *Reuters*, January 19, 1987.

21. The Council on Economic Priorities. Corporate Environmental Data Clearing House. *Shell. A Report on the Company's Environmental Policies and Practices* (New York, November 1991).

22. Shell International Petroleum Company (SIPC). Briefing note, December 1995.

23. J. P. Van Dessel, *The Environmental Situation in the Niger Delta, Nigeria*. Internal Position Paper for Greenpeace Nederland, February 1995, 12.

24. Greenpeace Nederland, *The Niger Delta*, 19.

25. SIPC briefing note, 1995.

26. Greenpeace Nederland, ibid.

27. Greenpeace Nederland, ibid., 43.

28. Ibid., 19.

29. Van Dessel, *Environmental Situation*, 15.

30. Khan, *Nigeria*, 40.

31. Khan, ibid., 32.

32. Greenpeace Nederland, *The Niger Delta*, 11-13.

33. IMF, International Financial Statistics, in Khan, *Nigeria*, 184.

34. Graf, *Nigerian State*, 218.

35. Khan, *Nigeria*, 136.

36. *Nigeria Today*, daily briefings of key political and economic developments in Nigeria (London, December 1996).

37. S. Ibi Ajayi, "An Economic Analysis of Capital Flight from Nigeria," World Bank policy working papers, 1992.

38. See Chapter Three of Khan, *Nigeria*, for details of oil companies and producing areas in the Niger Delta.

39. Ibid., 28.

40. Ibid., 22.

41. Greenpeace Nederland, *The Niger Delta*, 13.

42. Khan, *Nigeria*, 127.

43. See *Financial Times*, November 9, 1993. See also *Sunday Tribune* (Nigeria), June 11, 2000.

44. Project Underground, *Human Rights*, 5.

45. Minutes of meeting between Dr. Owens Wiwa, Ogoni Community Association U.K., and The Body Shop, London, November 23, 1995.

46. Project Underground, *Human Rights*, 9.

47. Project Underground, Interviews in the Niger Delta, April 12, 1997.

48. Eric Nickson, letter to Andrew Rowell and Paul Brown, November 6, 1996.

49. Project Underground, Interviews 1 through 4, April 12, 1997, in *Human Rights*.

50. Ibid.

51. Project Underground, Interviews 1 through 4 and letter from Police Constable No. 3, April 13, 1997.

52. Eric Nickson, letter to Rowell.

53. Project Underground, Interview No. 2, April 12, 1997.

54. Minutes of meeting between Dr. Owens Wiwa and The Body Shop, ibid.

55. Civil Liberties Organization, Environmental Rights Action, and others, *Shell in the Niger Delta: The Real Story*, November 1995, 9.

56. K. Bassee and B. Isreal, "A Detail Report of Our Arrest and Torture by Shell Staff/Armed Uniformed Men at Benson Beach in Oron—Oron Local Government Area, Akwa Ibom State," September 21, 1996. See also A. E. Robert, "A Statement on the Ogoni 19 Trial and Their Detention Conditions," August 2, 1996. See also Civil Liberties Organization, *Annual Report 1996* (Lagos, 1997), 200-202.

57. Claude Ake, "Shelling Nigeria," press statement, Port Harcourt, January 15, 1996.

58. See *PM News*, Lagos, "Shell's Formula for Vision 2010—Shonekan," October 23, 1996.

Chapter Four: A DYING LAND

1. Princess Irene Amangala, in the course of an interview with Alice Martins. "Oil, Oil Everywhere," *BBC Focus on Africa*, January-March 1995. Princess Amangala lives in Oloibiri, where Shell first struck oil in commercial quantities in 1956.

2. The World Bank, *Defining an Environmental Development Strategy for the Niger Delta*, 1995.

3. See Chapter Four of Nick Ashton-Jones with Susi Arnot and Oronto Douglas, *The ERA Handbook to the Niger Delta* (Environmental Rights Action, 1998).

4. J. P. Van Dessel, *Environmental Situation*, 8.

5. Ibid.

6. Niger Delta Wetlands Center, "Niger Delta Bio-diversity" (Port Harcourt, 1995).

7. Environmental Rights Action, "Destruction of the Delta" (Benin City, 1994).

8. World Bank, *Defining*.

9. Hans Magnus Enzensberger, in Roger Moody, "Courting Disaster," *Delta*, No. 2, November 1996.

10. Shell International Petroleum Company, briefing note, 1995.

11. Birinengi Idoniboye-Obu, "Resource Exploitation and Impact of Industrial Activities," in D. Fubara, ed., *The Endangered Environment of the Niger Delta: Constraints and Strategies*. An NGO Memorandum of the Rivers Chiefs and Peoples Conference, Port Harcourt, Nigeria, submitted to the World Conference of Indigenous Peoples on Environment and Development, Rio de Janeiro, June 1992, 59.

12. Quoted in Greenpeace International, *Shell-Shocked*, 11.

13. *Shell-Shocked*, 11, 32.

14. Ibid.

15. The Body Shop and Environmental Resources Management, *Review of Environmental Statements*, March 31, 1994.

16. Greenpeace Nederland press statement, November 10, 1996.

17. Shell International, *Shell and the Environment*, 1992, 5.

18. Idoniboye-Obu, "Resource Exploitation," 16.

19. Moffat and Linden, "Perception and Reality," 531.

20. Shell Petroleum Development Company (SPDC), *Annual Report 1996*, Lagos, May 1997.

21. World Bank, *Defining an Environmental Development Strategy for the Niger Delta*, Vol. 1, 49.

22. Greenpeace Nederland, *The Niger Delta*, 24.

23. David Moffat and Olof Linden, "Perception and Reality: Assessing Priorities for Sustainable Development in the Niger River Delta," *Ambio (Journal of the Swedish Academy of Sciences)*, Vol. 24, Nos. 7–8, December 1995, 532.

24. Greenpeace Nederland, *The Niger Delta*, 27. Shell has since commenced replacing some of these obsolete pipelines.

25. Van Dessel, *Environmental Situation*, 5.

26. Greenpeace Nederland, 26.

27. Ibid., 27.

28. See "Shell Is Biggest Global Warmer," Geoffrey Lean, *Independent on Sunday*, December 10, 1995. According to Lean, quoting figures supplied by the World Wide Fund for Nature (WWF), "Annually the flares emit 34 million tons of carbon dioxide, the main cause of global warming, while the oil fields vent about 12 million tons of methane, which has even more potent effect."

29. Greenpeace Nederland, 27.

30. *Independent on Sunday*, December 10, 1995.

31. Abdul Oroh, Executive Director of Civil Liberties Organization, personal communication, March 1995.

32. We relied on the *ERA Handbook to the Niger Delta* (1998), Van Dessel's internal position paper for Greenpeace Nederland, *The Environmental Situation in the Niger Delta* (February 1995), and the extensive fieldwork carried out in the Niger Delta by ERA volunteers to prepare this section.

33. Van Dessel, *Environmental Situation*, 14–15.

34. Moffat and Linden, "Perception and Reality," 527.

35. See particularly, Van Dessel, *Environmental Situation*, 16.

36. See Section V of Van Dessel, *Environmental Situation*.

37. Environmental Rights Action, *Shell and the Niger Delta: Between Rhetoric and Practice* (Benin City, June 1997).

38. Van Dessel, 22.

39. Ibid.

40. Ibid., 23.

41. Ibid.

42. Van Dessel, in *Price of Petrol, World in Action*, screened on ITV, May 13, 1996.

43. Van Dessel, *Environmental Situation*, 21.

44. Ibid., 17.

45. British Advertising Standards Authority Monthly, No. 62, July 10, 1996.

46. Claude Ake, in Channel 4 documentary, *The Drilling Fields*, Catma Films, May 23, 1994.

47. Van Dessel, 23.

48. Greenpeace Nederland, *The Niger Delta*, 26.

49. World Bank, *Defining*, Vol. 1; 58.

50. Khan, *Nigeria*, 164.

51. Environmental Rights Action, *Shell in Iko: The Story of Double Standards*, July 10, 1995, 25.

52. Khan, *Nigeria*, 160-62.

53. Van Dessel, 23.

54. Robert Corzine, "The Golden Egg Is Ready to Be Hatched," *Financial Times*, February 23, 1999. For details of the $3.8 billion NLNG project, see Greenpeace Nederland, *The Niger Delta*, 18.

55. Claude Ake, "Shelling Nigeria," January 15, 1996. See also D. Moffat and O. Linden, "Perception and Reality," 527-38. See also D. D. Ibiele, "Point-source Inputs of Petroleum Waste-water into the Niger Delta, Nigeria," in *The Science of the Total Environment*, 52, 1986; 233-38.

56. Ima Niboro, "Death on a Sea of Oil," *African Guardian*, October 4, 1993.

57. Ken Saro-Wiwa, "My Story," *Liberty, Quarterly Journal of the Civil Liberties Organization*, Vol. 7 No. 2, 1996.

58. Khan, *Nigeria*, 39-43.

59. Chief W. Nzidee, F. Yarika, N. Ndegwe, E. Kobani, O. Nalelo, Chief A. Ngei, and O. Ngofa, *Humble Petition of Complaint on Shell-BP Operations in Ogoni Division, Letter to His Excellency the Military Governor*, April 25, 1970.

60. A. Nedom and C. Kpakol, *Damages Done to Our Life-line by the Continued Presence of Shell-BP Company. Her Installations and Exploration of Crude Oil on Our Soil and Adequate Compensation Thereof.* Letter to the manager, Shell-BP Company of Nigeria, July 27, 1970.

61. Ken Saro-Wiwa, "My Story," *Liberty*.

62. P. Badom, *A Protest Presented to Representatives of the Shell-BP Development Company of Nigeria Ltd. by the Dere Youths Association; Against the Company's Lack of Interest in the Sufferings of Dere People, Which Sufferings Are Caused as a Result of the Company's Operations*, undated.

63. Jeremy Paul Spencer, Shell International Petroleum Company, statement to the UK Broadcasting Complaints Commission, May 29, 1996. Shell complaint regarding *Delta Force*, Catma films, which highlighted the shortcomings in the company's cleanup efforts in the Niger Delta. The Broadcasting Complaints Commission dismissed Shell's case. The World Council of Churches has also written, "Information provided above seriously calls into question Shell's version of the events surrounding the Ebubu oil spill." See World Council of Churches, Geneva, *Ogoni. The Struggle Continues*, December 1996, 34-37.

64. Quoted in Environmental Rights Action, *Environmental Human Rights. The Case Against Shell* (Benin City, 1996).

65. Chris McGreal, "Spilt Oil Brews Up a Political Storm," *Guardian*, August 11, 1993.

66. Greenpeace International, *Shell-Shocked*, 10.

67. R. W. Tookey, letter to Shelley Braithwaite, December 9, 1992.

68. Quoted in Andrew Rowell, *Green Backlash: Global Subversion of the Environment Movement* (London: Routledge, 1996), 311–12.

69. Claude Ake, interview with Catma Films, January 1994.

70. Nick Ashton-Jones in *Green Backlash*, 291.

71. Ken Saro-Wiwa, *Shell in Ogoni and the Niger Delta*, EMIROAF Report (Port Harcourt, 1992).

72. Shell officials, in Greenpeace International, *Shell-Shocked*, 10.

73. Van Dessel, *Environmental Situation*, 23.

74. Project Underground, *Human Rights*.

75. Idoniboye-Obu, "Resources Exploitation."

76. Obasi Ogbonnaya, "The Ghost Shell of a Billion Dollar Enclave," *Guardian*, Lagos, April 3, 1994.

77. Ibid.

78. Chief Famous Alawari of Okoroba in conversation with British environmentalist Nick Ashton-Jones. See ERA Report, *Environmental Human Rights*, 1.

79. Alice Martins, "Oil, Oil Everywhere," *BBC Focus on Africa*, January–March 1995, 10.

80. Environmental Rights Action, *Environmental Human Rights*.

81. Alice Martins, based on an interview with Chief Richard Ekalabo of Okoroba village, "Oil, Oil Everywhere," 12. The quote is a translation of Ekalabo's comments to the BBC journalist.

82. Environmental Rights Action, *Environmental Human Rights.*

83. Martins, ibid.

84. Chief Famous Alawari, in Obasi Ogbonnaya, "Ghost Shell."

85. This is a summary of Environmental Rights Action, *Shell in Iko: The Story of Double Standards,* July 1995.

86. Ibid., 13.

87. Ibid., 5.

88. Edem Esara, research project conducted in Iko, cited in Environmental Rights Action, *Shell in Iko*, 13–18.

89. Greenpeace Nederland, 28.

90. Edem Esara, research project, in *Shell in Iko.*

91. Environmental Rights Action, *Shell in Iko,* 14.

92. Ibid.

93. Ibid., 25.

94. Shell Petroleum Development Company figures, quoted in Greenpeace Nederland, 25.

95. Ibid.

96. Van Dessel, *Environmental Situation,* 24. See also Greenpeace Nederland, 25.

97. World Bank, *Defining,* Vol. 1; 48.

98. Greenpeace Nederland, 25.

99. Moffat and Linden, "Perception and Reality," 532.

100. Claude Ake, "Shelling Nigeria."

101. Environmental Rights Action briefing paper, July 1997.

102. Ake, "Shelling Nigeria." (PPM—Parts Per Million)

103. Greenpeace Nederland, 25.

104. World Bank, *Defining,* Vol. 1; 51.

105. Greenpeace Nederland, *The Niger Delta,* 25.

106. Ibid., 12.

107. World Bank News, "IFC Pulls Out of Nigeria LNG Project," November 16, 1995.

108. Shell Venster, 1996, 12-13.

109. SGS Environment Ltd., *Nigeria LNG Project, Combined Executive Summary* (Liverpool, March 1995).

110. Nick Ashton-Jones, letter to Osman Shahenshah, Oil and Gas Division, International Finance Corporation, Washington, July 11, 1995.

111. SGS Environment Ltd., *Nigeria LNG Project.*

112. Ibid., 13-17.

113. Nick Ashton-Jones, letter to Shahenshah.

114. Phil Smith, Aquatic Environmental Consultants, *Review of the Environmental Statements Prepared for Nigeria LNG Ltd. by SGS Environment Ltd.*, December 8, 1995.

115. Figures quoted in SGS, *Nigeria LNG Project.*

116. SGS Environment Ltd., Nigeria, Combined Executive Summary, Liverpool, March 1995.

117. Mathews Tostevin, "Protesters Renew Nigeria Energy Worries," Reuters, September 23, 1999.

118. IRIN West Africa Update, "Nigeria: Bonny Chiefs Accuse Oil Firms," July 6, 1999.

119: Joseph Ollor Obari, "President, in Bonny, Alerts of Threat to Economy," *Guardian*, Lagos, September 27, 1999.

120. Ibid.

121. Humphrey Bekaren, "Shell's Cock and Bull Story," *Guardian*, Lagos, undated.

122. Shelley Braithwaite, "Leaks in the System: An Environmental and Social Investigation into Shell's Nigerian Operations Relative to European Standards." Unpublished Master of Science dissertation, submitted to Brunel University, England, April 1999.

123. Ibid.

124. Michael Fleshman, "Report from Nigeria 2," The Africa Fund, New York, June 17, 1999.

125. Environmental Rights Action, *Shell and the Niger Delta.*

126. Ken Saro-Wiwa, speech delivered on the occasion of a Nigerian government ministerial visit to Ogoni in January 1993.

Chapter Five: WHERE VULTURES FEAST

1. Greenpeace International, *Shell-Shocked*. This statement by a senior BP petroleum engineer was taken from Rowell's report for Greenpeace.

2. Chris McGreal, "Nigeria's Golden Egg Left to Rot," *Guardian*, London, December 6, 1995.

3. Ibid.

4. Shell International Petroleum Company, *Shell in Nigeria* (London, November 1995).

5. Quoted in Sam Olukoya, "Why They Seethe," *Newswatch*, December 18, 1995.

6. Shell Petroleum Development Company, *Community 1996*, Lagos.

7. Shell International, briefing note, November 1995.

8. Greenpeace Nederland, *The Niger Delta*, 18.

9. The Shell Transport and Trading Plc, *Annual Report 1991*, 6. See also D. Buckman, "West African Prospects Still Look Good," *Petroleum Review*, November 1993.

10. Shell International, *Shell in Nigeria*.

11. ShellVenster, January–February, 1996; 20–23.

12. Greenpeace Nederland, *The Niger Delta*, 13.

13. See Khan, *Nigeria*, 81-85, for details of MOUs.

14. World Petroleum Laws, Petroconsultants Economics Division, *Nigeria*, March 1994.

15. *Petroleum Economist*, Special Supplement, Energy Law, July 1992.

16. Greenpeace Nederland, *The Niger Delta*, 43.

17. Khan, *Nigeria*, 78. See also Van Dessel, *Environmental Situation*, 15, for SNEPCO's activities in the Gongola Basin, where Nigeria's famous Yankari Game Reserve is located.

18. Khan, *Nigeria,* 78.

19. See Ademola Adedoyin, "Why He Struck," *The Week*, September 9, 1996.

20. Ibid.

21. Ibid.

22. Pascal Nwigwe, "Shell Drags House Staff to Court over Staff," *ThisDay*, February 20, 2000.

23. Ademola Adedoyin, ibid.

24. Ibid.

25. *Petroleum Economist*, March 1992.

26. Jedrzeg George Frynas, "Political Instability and Business: Focus on Shell in Nigeria," *Third World Quarterly*, Vol. 19 No. 3, 1998; 457–78.

27. Greenpeace Nederland, *The Niger Delta*, 13.

28. Adedoyin, "Why He Struck."

29. "Eyeball to Eyeball," *Newswatch*, September 16, 1996.

30. Greenpeace Nederland, *The Niger Delta*, 43.

31. Robert Corzine, "The Golden Egg Is Ready to Be Hatched," *Financial Times*, February 23, 1999. Nick Ashton-Jones also studied some of Shell's community development projects in Nembe, 1990–93, and concluded that they were inflated. See ERA Monitor Report No. 3, *Shell and Community Support in the Niger Delta*, January 1998.

32. Shell Petroleum Development Company, *The Ogoni Issue*, 1995.

33. Ken Saro-Wiwa, *Shell in Ogoni and the Niger Delta*, EMIROAF Report, December 8, 1992.

34. Ibid.

35. Shell Petroleum, *Ogoni*.

36. Environmental Rights Action, *Briefing on Ekeremor-Zion*, December 1997.

37. *Nigeria Weekly Law Report*, 1995 (3), Part 382; 148.

38. Ima Niboro, "Off to a Staggering Start," *Tell*, January 6, 1997; p.17.

39. Environmental Rights Action, *Shell in Iko*, 9.

40. Anietie Usen, "Mission to Iko," *Newswatch*, December 23, 1985; 4.

41. Shell Petroleum, *Shell East Community Assistance Projects: 1986–1993*, October 1995.

42. Civil Liberties Organization and others, *Shell in the Niger Delta*.

43. Ogbonnaya, "Ghost Shell."

44. Ronnie Siakor, *A Study of Some of Shell's Community Development Projects*, Prepared for Living Earth, London, April 1996.

45. Sunday Times Insight Team, "Shell Purges Top Staff in Nigeria," *Sunday Times*, London, December 17, 1995.

46. Ibid.

47. Ibid.

48. Ibid.

49. Shell International, *Briefing Notes*, 1995.

50. Rowell, *Green Backlash*, 289.

51. *Nigeria Weekly Law Report*, ibid.

52. Nick Ashton-Jones, *Oil Production in the Niger Delta*, confidential memorandum, January 1996.

53. Rowell, *Green Backlash*, 27.

54. Nick Ashton-Jones, Oronto Douglas, and Uche Onyeagucha, *The Human Habitat of Port Harcourt*, Report for Pro Natura International, January 1995.

55. Quoted in Moffat and Linden, "Perception and Reality," 536.

56. Shell International, *Shell in Nigeria*, December 1995.

57. Barrows, *IPI Data Service Africa*, No. 47, *Nigeria*, 1986, 17. See also Khan, *Nigeria*, 19.

58. Khan, Nigeria, 33.

59. World Bank, *Defining*, Vol. 1, 48–49.

60. Moffat Ekoriko, "More Evidence of Oil Devastation," *Africa Today*, September–October 1996.

61. Environmental Rights Action, *The Silent Spill*, June 14, 1995.

62. "Mobil Poisons the Seas, Endangers Communities," *Niger Delta Alert*, No. 1, January 1998.

63. Bruce Powell, the late Canadian environmentalist and scholar who lived and worked in the Niger Delta, made this statement in the course of an interview with Human Rights Watch on June 20, 1998. See Human Rights Watch, *The Price of Oil. Corporate Responsibility and Human Rights Violations in Nigeria's Oil Producing Communities* (Human Rights Watch, New York, January 1999), 71. For details of the court case against Mobil in Nigeria, see *Delta*, No. 2, November 1996, 10.

64. Human Rights Watch/Africa, *Ogoni Crisis*, 32.

65. See Human Rights Watch/Africa, *Ogoni Crisis*, 32–35, for details of Professor Chinwah's ordeal.

66. See Nick Ashton-Jones, *Renewable Natural Resources in Southern Nigeria and Agroforestry*, ERA Monitor Report 4, February 1998.

67. Ashton-Jones, *Oil Production in the Niger Delta.*

Chapter Six: AMBUSH IN THE NIGHT

1. "Plight of the Ogoni," *Newsweek*, September 20, 1993.

2. Ken Saro-Wiwa, "My Story." Full text of Saro-Wiwa's statement to the Ogoni Civil Disturbances Tribunal, in *Ogoni: Trials and Travails* (Lagos: Civil Liberties Organization, 1996), 41.

3. See *Ogoni Bill of Rights*.

4. Saro-Wiwa, *Trials*, 41.

5. Diana Wiwa, "The Role of Women in the Struggle for Environmental Justice in Ogoni," *Delta*, No. 3, October 1997, 11.

6. Saro-Wiwa, *Trials*, 45.

7. World Council of Churches, *Ogoni. The Struggle Continues* (Geneva: December 1996), 90.

8. Human Rights Watch/Africa, *The Ogoni Crisis*, 10.

9. Shell Petroleum Development Company, *Meeting at Central Offices on Community Relations and Environment, 15–16 February in London, 18 February in The Hague*, 1993, draft minutes.

10. Saro-Wiwa, *Trials*, 46.

11. Shell Petroleum, *The Ogoni Issue.* The World Council of Churches has, however, argued that it is not true that Shell oil production in Ogoni ceased completely in mid-1993, as gas was still being flared in the Ebubu flow station in 1995. See World Council of Churches, *Ogoni*, 33.

12. Saro-Wiwa, *Trials*, 47.

13. Ibid., 48. This sequence of events has since been confirmed by the Civil Liberties Organization, which has been monitoring developments in Ogoniland since 1992. Personal comments, Abdul Oroh, Executive Director of Civil Liberties Organization, to Ike Okonta, July 1, 1997.

14. Saro-Wiwa, *Trials*, 49.

15. Ibid., 50.

16. Irene Bloemink to Ike Okonta and Oronto Douglas, personal communication, July 1997.

17. World Council of Churches, *Ogoni*, 91.

18. Ibid.

19. T. Owolabi, "Genocide in Ogoni," *Sunday Tribune*, October 1996, 21. Also see Saro-Wiwa, *Report to Ogoni Leaders Meeting at Bori*, 1993.

20. See Colonel Okuntimo's statement in *Sunday Times*, London, December 17, 1995. See also Charles Danwi and Nayone Akpa's affidavit in Birnbaum, *Nigeria. Fundamental Rights Denied*, 36, and Appendix 10: summary of affidavits alleging bribery.

21. Andy Rowell, "Sleeping with the Enemy," *Village Voice*, New York, January 23, 1996. See also Owens Wiwa, *Minutes of Meeting*. See also Saro-Wiwa, "Statement to the Ogoni Civil Disturbances Tribunal," Port Harcourt, November 1, 1995.

22. Saro-Wiwa, *Trials*, 56.

23. Ake in *The Drilling Fields*, Catma Films, May 23, 1994.

24. Human Rights Watch/Africa, *The Ogoni Crisis*, 12.

25. Ibid., 13.

26. Saro-Wiwa, quoted in publicity poster of the New Nigeria Forum, London, July 1995.

27. Saro-Wiwa, *Report to Ogoni Elders*.

28. Polly Ghazi, "Shell Letters Show Nigeria Army Links," *Observer*, December 17, 1995. See also Ashton-Jones, *Shell Oil in Nigeria*, August 1994.

29. Saro-Wiwa, *Trials*, 58.

30. See Birnbaum, *Nigeria*, 11, for Dr. Leton's position.

31. Lt. Colonel Paul Okuntimo, Rivers State Internal Security Task Force, memorandum to Colonel Dauda Komo, military administrator of Rivers State. Government House, Port Harcourt, May 12, 1994.

32. G. B. Leton, statement to the Ogoni Civil Disturbances Tribunal.

33. Saro-Wiwa, *Trials*, 95.

34. See Chief Kemte Giadom's testimony to the Ogoni Civil Disturbances Tribunal, in Birnbaum, *Nigeria*, Appendix 9, 104.

35. See the testimony of Assistant Superintendent of Police Stephen Hasso in Birnbaum, *Nigeria*, 104.

36. Press conference, Port Harcourt, Nigerian Television Authority, May 22, 1994.

37. Ake, "A People Endangered by Oil," *Guardian*, Lagos, July 18, 1994.

38. *Guardian*, Lagos, editorial comment, July 18, 1994.

39. Okuntimo, in *Trials*.

40. Sylvester Olumhense, Oronto Douglas, and Uche Onyeagucha, "The 1994 Massacre," in *Trials*, 28-39.

41. Shell Petroleum, *Ogoni Issue*.

42. Ghazi, *Observer*, November 19, 1995.

43. See preface of Birnbaum, *Nigeria*.

44. Shell International, *Verdict on the Trial of Ken Saro-Wiwa*, October 1995.

45. Rowell, "Shell Cracks: Petroleum Company Acknowledges Payments to Nigeria's Murderous Military," *Village Voice*, New York, December 17, 1996.

46. Oronto Douglas, "Ogoni: Four Days of Brutality and Torture," *Liberty*, May–August 1994.

47. Insight Team, "Shell Purges," *Sunday Times*.

48. Human Rights Watch/Africa, *The Ogoni Crisis*, 38. See also Ledum Mitee in "Shell Purges."

49. "Shell Purges," *Sunday Times*.

50. Oronto Douglas, "Ogoni: Four Days." Nick Ashton-Jones has also confirmed Oronto Douglas's account of the incident at Bori Camp in Port Harcourt and subsequent events at the State Security Service Headquarters. See Nicholas Ashton-Jones's letter to Michael Birnbaum in Birnbaum, *Nigeria*, 95.

51. Polly Ghazi and Cameron Duodu, "How Shell Tried to Buy Berettas for Nigeria," *Observer*, February 11, 1996.

52. Ibid.

53. Ibid.

54. Environmental Rights Action, *Shell in Iko*.

55. J. R. Udofia, Shell Petroleum Development Company, letter to the Commissioner of Police, Rivers State, October 29, 1990.

56. Justice O. Inko-Tariah and others, *Commission of Inquiry to the Causes and Circumstances of the Disturbance that Occurred at Umuechem in the Etche Local Government Area of Rivers State*, Port Harcourt, 1990.

57. Constitutional Rights Project, "Time to Talk," *Constitutional Rights Journal*, Vol. 3 No. 8, October–December 1993, 8.

58. Ogbonnaya, "Ghost Shell."

59. Human Rights Watch/Africa, *The Ogoni Crisis*, 35.

60. Letter from the Rumuobiokani Community to the Military Administrator of Rivers State and the General Manager (East) Shell Petroleum Development Company, February 24, 1994.

61. Human Rights Watch/Africa interviews, Port Harcourt, March 2 and 3, 1995, in *The Ogoni Crisis*.

62. Human Rights Watch/Africa interviews, March 3, 1995.

63. Movement for the Survival of Ijaw Ethnic Nationality (MOSIEND) press release, March 14, 1994.

64. Martin, "Oil, Oil Everywhere."

65. Human Rights Watch/Africa, *The Ogoni Crisis*, 32.

66. Iyobosa Uwugiaren, "More Trouble for Shell, Others," *PM News*, October 20, 1997.

67. Movement for the Survival of the Ogoni People, press release, December 6, 1996, Port Harcourt.

68. Environmental Rights Action, "Don't Militarize Bayelsa," press statement, August 12, 1997.

69. Radio Nigeria, Kaduna, announced on December 5, 1998, that the military government had finalized plans to build a naval base in Baylesa State.

70. See *The Kaiama Declaration, Resolutions of the December 11, 1998, All Ijaw Youths Conference Held in the Niger Delta, Nigeria,* published for the Ijaw Youth Council by the Ijaw Council for Human Rights (Port Harcourt, March 1999).

71. See articles iv-vi of the Kaiama Declaration.

72. See the ten-point resolution of the Kaiama Declaration.

73. Felix Tuodolor, president of the Ijo Youth Council, in an interview with Ike Okonta, January 4, 1999.

74. Patterson Ogon, coordinator of the Ijo Council for Human Rights, in an interview with Ike Okonta, January 4, 1999. See also *Ogele,* bulletin of the Ijo Youth Council, December 30, 1998.

75. Reuters news report, January 1, 1999.

76. Ijo Council of Human Rights, *Barrels of Blood: Human Rights Atrocities Following the Kaiama Declaration*, June 1999. A Human Rights Watch team that visited the area after the Yenagoa killings also confirmed the number of deaths in the first encounter between the youths and government troops on the morning of December 30. See Human Rights Watch, *Crackdown in the Niger Delta*, May 1999, 6.

77. Human Rights Watch, *Crackdown in the Niger Delta*, 7.

78. *PM News*, Lagos, January 1, 1999.

79. Blessing Ajoko, Interim Secretary, Ogbia Youth Vanguard, in an interview with *Niger Delta Alert* (monthly publication of Delta Information Service), April 1999.

80. Anemeyeseigha Brisibe, coordinator of Niger Delta Women for Justice, Port Harcourt, in an interview with Ike Okonta, January 12, 1999.

81. Adaka Isaac Boro later joined the federal forces when civil war broke out and was killed in late 1969, just when the war was about to end. See John De St. Jorre, *The Nigerian Civil War* (London: Hodder and Stoughton, 1972), 152.

82. Human Rights Watch, *Crackdown in the Niger Delta*, 10.

83. Ijo Council of Human Rights, "Dirty War," excerpt from *Barrels of Blood: Human Rights Atrocities Following the Kaiama Declaration*, in *Eraction*, quarterly journal of Environmental Rights Action, January–March 1999, 21–23.

84. Human Rights Watch, *Crackdown in the Niger Delta*, 6.

85. Ibid., 11.

86. Sola Omole, a senior manager with Chevron Nigeria, accepted in an interview broadcast on Pacifica Radio that his company's management had authorized the invitation of the navy to the

platform and also provided the helicopters to ferry them to the scene. See Amy Goodman and Jeremy Scahill, "Drilling and Killing: Chevron and Nigeria's Oil Dictatorship," transcript of broadcast on *Democracy Now* program, Pacifica Radio, New York, October 1, 1998. See also Human Rights Watch, *The Price of Oil*, 151.

87. Environmental Rights Action, "Chevron's Commando Raid," June 1998.

88. Patterson Ogon of Ijaw Council for Human Rights in an interview with Ike Okonta, June 6, 1999.

89. Human Rights Watch visited Opia and Ikenyan communities in February 1999 and was able to establish, through eyewitness accounts, that vehicles (including a helicopter) contracted to Chevron were used in these attacks. See Human Rights Watch, *Crackdown in the Niger Delta*, 14.

90. Human Rights Watch, *Crackdown in the Niger Delta*, 14.

91. Olusola A. Omole, General Manager Public Affairs, Chevron Nigeria Ltd., in a letter to Opia and Ikenyan communities, dated January 22, 1999, quoted in Human Rights Watch, *Crackdown in the Niger Delta*, 15–16.

92. Patterson Ogon, in an interview with Ike Okonta, June 6, 1999.

93. Mike Libby, speaking for Chevron's U.S. headquarters on the program *Democracy Now*, Pacifica Radio, New York, February 24, 1999, and quoted in Human Rights Watch, *Crackdown in the Niger Delta*, 17.

94. See Human Rights Watch, *The Price of Oil*, 151.

95. Tape of Radio Pacifica's *Democracy Now*, April 30, 1999, quoted in Human Rights Watch, *Crackdown in the Niger Delta*, 17.

96. Olusola A. Omole, letter to Opia and Ikenyan communities.

97. Bassey Udo, "Nigeria reaps N1.7b windfall as oil price rises to $18.69," *Post Express*, Lagos, July 15, 1999.

98. Saro-Wiwa, *Trials*, 100.

Chapter Seven: A GAME FOR SPIN DOCTORS

1. Brian Anderson, press statement, Shell Petroleum Development Company Ltd., November 8, 1995, Lagos.

2. Ibid.

3. Charles Suanu Danwi, a principal prosecution witness in the trial of Ken Saro-Wiwa and the other MOSOP activists, later said under oath he was bribed by Shell and the military junta to testify against Ken Saro-Wiwa. Danwi's exact words: "Most of the people identify [*sic*] never go to Giokoo or take part in the murder. But they wanted me to make sure I made a statement that will involve Ken, Mitee and MOSOP officials so that they will kill them." See Birnbaum, *Nigeria*.

4. Anne McElvoy, "The Moral Daze," *Spectator*, January 13, 1996.

5. Ibid.

6. Minutes of meeting, Shell Petroleum Development Company.

7. Ibid.

8. Ibid.

9. Moody, *The Gulliver File*, 703-704.

10. Kenny Bruno, in Rowell, *Green Backlash*, 101.

11. "The PR Industry's Top 15 Greenwashers, Based on O'Dwyer's Directory of PR Firms and Interviews with Hill and Knowlton," *PR Watch*, Vol. 21 No. 1, First Quarter, 1995, 4.

12. J. C. Stauber, "Going . . . going . . . green," *PR Watch*, Vol. 1 No. 3, 1994, 2.

13. Rowell, *Backlash*, 112.

14. C. Fay, Interview on BBC Radio 4, One O'clock News, September 8, 1996.

15. See "A Slice of Good Sense," *Guardian*, February 2, 1998.

16. Rowell, "Unloveable Shell, the Goddess of Oil," *Guardian*, November 15, 1997.

17. See *Sunday Times*, "Shell Purges."

18. The Right Livelihood Award Foundation, "1994 Right Livelihood Awards Stress Importance of Children, Spiritual Values and Indigenous Cultures," press release, November 10, 1994.

19. Record of meeting held between the Nigerian High Commissioner to Britain, Alhaji Abubakar Alhaji, and four senior officials of Shell International Petroleum Company at Shell Centre, London, March 16, 1995.

20. Ibid.

21. A. Garret, "Sponsorship: Business Buys Its Way to a Greener Image," *Independent on Sunday*, November 10, 1991.

22. Rowell, *Backlash*, 112.

23. Shell International Petroleum Company, *The Niger Delta Environmental Survey*, November 1995.

24. Richard Dowden, "'Green' Shell Shares Sold in Protest at Spills," *Independent*, October 24, 1994.

25. Saro-Wiwa, private letter to Mr. Gamaliel Onosode, Chairman of the Steering Committee of the Niger Delta Environmental Survey (NDES), August 27, 1995.

26. Minutes by a potential contractor of a meeting held with Shell Environmental Division, quoted in Rowell, *Backlash*, 312.

27. G. O. Onosode, keynote address during a Niger Delta Environmental Survey consultative meeting with stakeholders in Delta State, November 28, 1995.

28. See Human Rights Watch, *The Price of Oil*, 88. See also Polly Ghazi, *Observer*, December 10, 1995.

29. *Delta*, No. 1, April 1996.

30. Claude Ake, letter of resignation to Gamaliel Onosode, Chairman of the Steering Committee of Niger Delta Environmental Survey, November 15, 1995.

31. Shell International, Shell briefing notes, November 1995.

32. Onosode, keynote address.

33. Shell Petroleum Development Company, press statement by Brian Anderson, Lagos, November 8, 1995.

34. John Drake, Shell South Africa (Pty) Ltd., letter to President Nelson Mandela, November 24, 1995.

35. Dominic Midgley, "The Man Who Fooled the World," *Punch*, London, February 1998.

36. Andrew Neil, "Saint Ken—Violent Hero of the Gullible," *Mail*, November 16, 1995.

37. "Free Trips for a Free Press," *Delta*, No. 2, November 1996, 9.

38. Ibid.

39. Ibid.

40. Environmental Rights Action, minutes of meeting between Environmental Rights Action, Swedish Society for Nature Conservation, Shell Sweden, Shell International London, and Friends of the Earth, Sweden, October 1996.

41. John Mcmanus, "Shell Damage Limitation Tour Offers No Easy Answers," *Irish Times*, February 2, 1996.

42. Quoted in *Guardian*, Lagos, November 18, 1996.

43. O. Isralson, "Washington Lawmakers: Under the Influence; U.S. Lobbyists Have the World's Interest at Heart," *World Paper*, February 1990, 10. B. Vora, "Beyond Hiring a Lobbying Firm, the Ethnic Newswatch," *News India*, March 4, 1994, Vol. 24 No. 9; 54.

44. *Crisis in Ogoniland: How Saro-Wiwa Turned MOSOP into a Gestapo*. No publisher. Distributed by Nigeria's information minister at Oxford, May 1, 1995.

45. Rowell, *Green Backlash*, 314.

46. Embassy of the Federal Republic of Nigeria, Dublin, Ireland, *Truth of the Matter*, booklet printed by the embassy and sent to Trócaire and other NGOs campaigning against Shell in Europe.

47. John J. Barry, Shell International Petroleum Company Ltd., letter to J. Kilcollen, Director, Trócaire, County Dublin, Ireland, March 6, 1995.

48. "Nigeria Foaming," *The Economist*, November 18, 1995.

49. *The Economist*, June 16, 1900.

50. Shell International, "Shell Reaffirms Support for Human Rights and Fair Trial," news release, London, January 30, 1996.

51. Civil Liberties Organization, *Annual Report 1996*, 202–203.

52. "Shell Purges," *Sunday Times*.

53. Shell Petroleum Development Company Ltd., Circular No. OL 1312, July 19, 1996, Lagos.

54. Greenpeace Nederland, *The Niger Delta*, 49.

55. Summaries of Shell's complaint to the United Kingdom Broadcasting Complaints Commission regarding the Catma Film, *Delta Force*.

56. Jeremy Paul Spencer, Safety and Environmental Affairs Manager, Shell U.K. Expro, *Statement to the Broadcasting Complaints Commission*, May 26, 1996.

57. Channel 4 Television, "Shell Climbs Down Over Ogoniland Complaint," press release, December 18, 1996.

58. See *Guardian*, Lagos, October 17, 1996.

59. *Delta*, "Shell Slips in Through the Backdoor," No. 2, November 1996. See also *Sunray*, Port Harcourt, July 29, 1996.

60. Brian Anderson, in an interview in *ThisDay*, Lagos, November 10, 1996.

61. Movement for the Survival of the Ogoni People, press release, Port Harcourt, January 6, 1997.

62. John Barry, letter to Kilcollen.

63. Shell International, *Shell and Nigeria*, May 1995.

64. "Shell Purges," *Sunday Times*.

65. Human Rights Watch/Africa, *The Ogoni Crisis*, 38.

66. "Shell Cracks," Rowell.

67. Shell Petroleum Development Company, press statement by Brian Anderson, November 8, 1995.

68. Birnbaum, *Nigeria*, Appendix 10; 117.

69. Civil Liberties Organization and others, *Shell in the Niger Delta*.

70. Shell Petroleum Development Company, *The Environment*, May 1995.

71. P. Clothier and E. O'Connor, "Pollution Warnings 'Ignored by Shell,'" *Guardian*, May 13, 1996.

72. See Rowell, *Green Backlash*, 293.

73. C. A. J. Herkstroter, letter to Teo Wams of Milieudefensie, Amsterdam, October 24, 1996.

74. Ghazi, *Observer*, December 10, 1995.

75. Shell Petroleum Development Company, press statement, November 8, 1995.

76. Shell Petroleum Development Company, *Community Development*, October 1995.

77. "Shell Purges," *Sunday Times*.

78. Nnaemeka Achebe, speaking to John McManus, *Irish Times*, December 2, 1996.

79. See *Ogoni Bill of Rights*.

80. Shell Petroleum Development Company, press statement, November 8, 1995.

81. Shell Petroleum Development Company, *The Ogoni Issue*, January 1995.

82. Nnaemeka Achebe, speaking to McManus.

83. Shell Petroleum Development Company, *Shell in Nigeria*, January 1995.

84. *Delta*, No. 2, November 1996, 21.

85. Royal Dutch Shell advertisement, *Landelijke Nederlandse Dagbladen*, November 21, 1995.

86. Shell Petroleum Development Company, *Ogoni Issue*.

87. *Advertising Standards Authority Monthly*, No. 62, July 1996.

88. Shell Petroleum Development Company, *The Environment*, May 1995.

89. Greenpeace Nederland, *The Niger Delta*, 49.

90. Environmental Rights Action, *Shell in Nigeria: Public Relations*.

91. Ibid.

92. Shell Petroleum Development Company, *Environment 1996*.

93. Wolfgang Mai, *Observations After a Visit to the Oil Producing Areas of Nigeria* (October 25 to November 9, 1997), November 25, 1997, 8.

94. See Human Rights Watch, *The Price of Oil*, 182.

95. Shell, *Profits and Principles*, 1998, 2.

96. Richard Swift, "Mindgames," *New Internationalist*, July 1999.

97. See Shell-sponsored guide, "Responsible Business in the Global Economy," *Financial Times*, June 1999.

98. *The Economist*, London, June 19–25, 1999.

99. Michael Fleshman, "Report from Nigeria 2," *Nigeria Transition Watch Number 9*, June 17, 1999, New York.

100. Ibid.

Chapter Eight: HEALING THE WOUND

1. Cheikh Anta Diop, *Civilization or Barbarism. An Authentic Anthropology* (New York: Lawrence Hill Books,1991), 375.

2. Barry Hillenbrand, "Seeing the Light in Nigeria," *Time*, November 25, 1996, 69.

3. Moffat and Linden, "Perception and Reality," 534.

4. Ibid., 532.

5. Ibid., 534.

6. Douglas Ashton-Jones, and Onyeagucha, *The Human Habitat*.

7. Ibid.

8. Ibid.

9. Ashton-Jones and Douglas, *The Human Ecosystem of the Niger Delta. A Preliminary Baseline Ecological Survey Participatory Survey of the Niger Delta*, May 1994.

10. See Moffat and Linden, "Perception and Reality," 531.

11. Ashton-Jones and Douglas, ibid.

12. Moffat and Linden, "Perception and Reality," 528.

13. Van Dessel, *Environmental Situation*, 29.

14. Ibid.

15. Ashton-Jones, "Delta Remembers."

16. Moody, "Courting Disaster," *Delta*, No. 2, November 1996.

17. See first page of Rowell, *Green Backlash*.

18. Rowell, ibid., 373.

19. Vandana Shiva in Rowell, ibid., 374.

20. Barry Hillenbrand, ibid.

21. See Chapter 15 of *ERA Handbook to the Niger Delta*.

22. Ibid.

23. See Appendices for details of ERA and its activities.

24. Anne McElvoy, "The Moral Daze."

25. Resolution of Joint Committee on Foreign Affairs, Leinstre House, Dublin 2, January 17, 1996.

26. Tajudeen Abdul Raheem, *Twenty Questions and Answers on Sanctions Against Nigeria*, November 1995, unpublished.

27. Trócaire, *The Ogoni Tragedy*. Trócaire's response and recommendations for action, Parliamentary briefing.

28. Khan, *Nigeria*, 117.

29. Dan Atkinson, "OPEC Over a Barrel," *Guardian*, April 1, 1998.

30. Atkinson, ibid.

31. Khan, *Nigeria*, 120. Also see "Nigeria Creates New Police Unit to Guard Oil Pipelines," Dow Jones International News, March 22, 2000.

32. See "The Samson Option," *Nigeria Now*, New Nigeria Forum, London, Vol. 7, June–December 1997, 7.

33. "Clinton Softens Stance Toward Abacha," *Nigeria Today*, London, March 30, 1998. See also Charles R. Babcock and Susan Schmidt, "Voters Group Donor Got DNC Perk; Man with Nigeria Ties Was at Clinton Dinner," *The Washington Post*, November 22, 1997.

34. *Nigeria Now*, ibid. See also Ian Black, "UK Resists Call to Oust Nigeria," *Guardian*, September 11, 1997.

35. Quoted in David Rowan, "Meet the New World Government," *Guardian*, February 13, 1998.

36. Simon Caulkin, "Amnesty and WWF Take a Crack at Shell," *The Observer*, London, May 11, 1997.

37. Geoffrey Gibbs, "Illness in Wake of Oil Spill," *Guardian*, London, December 4, 1996.

Appendix: JUSTICE ON TRIAL

1. *Delta*, No.2, November 1996. See also Human Rights Watch, *The Price of Oil*, 169.

2. Shell International: Environmental Clinic Nigeria/Shell Team. Memorandum prepared in anticipation of litigation, February 26, 1996.

3. We find the style used by Professor D. J. Hughes in discussing the "environment" very useful. Professor Hughes divides the environment into "three"—land, air, and water. See David J. Hughes,

Environmental Law (Butterworths, London, 3rd edition, 1996), 1-541. However, this does leave out the spiritual and religious dimension, which is a significant anchor of the Niger Delta environment.

4. By this we mean "existing legal rules which have partial or total bearing on the environment of Nigeria. Such rules could spell out the use, protection, preservation or conservation of a thing or an entire ecosystem." Oronto Douglas, *An Overview of Environmental Protection Law in the Niger Delta*. Dissertation submitted in partial fulfillment of an M.A. degree, DMU, Leicester, 1996.

5. For a detailed discussion on sources of Nigerian law see Professor Niki Tobi, *Sources of Nigerian Law*, Lagos, 1996.

6. Interpretation Act 1964 (No. 1 of 1964) S. 28; Law (Miscellaneous Provisions): Lagos Laws 1973 Cap 65 and the High Court Law of Lagos 1973 Cap 52; Law of England (application) Law; Western Region of Nigeria Laws 1959 Cap 60. Also see The High Court Law of the Northern States NN Laws 1963 Cap 49 SS 28, 29, and 35.

7. Applicable laws (Miscellaneous Provisions) Edict (1988 No. 10). Published in the *Gazette* of Rivers State 1989, No 20. Now also applicable to the newly created Bayelsa State. The law is an offshoot from the former Eastern Nigeria Law EN Laws 1963 Cap 61. See particularly Section 15.

8. The true story of who benefits more, Britain or Nigeria, is beyond a report of this nature. See Walter Rodney, *How Europe Underdeveloped Africa*, Nigeria edition (Enuju: Ikenga Publishers, 1982).

9. It is suggested that the unwieldy nature of the Rivers State edict gives room for speculation as to which "Ireland" it means—North or the Republic—since even the Republic of Ireland has been bitten by the bug of the Common Law of England. The speculation will not go away. What is clear, however, is that the enactment means business: the application of laws of the former colonial master. One such Common Law waiting to be applied is the recent case of *Cambridge Water Co. Ltd.* v. *Eastern Counties Leather Plc* (1994), House of Lords.

10. The tiers are the federal, state, and local governments.

11. S.I. 1954, No. 1146.

12. The Designation of Ordinances Act No. 57 of 1961.

13. See Decree 107 of 1993, which vests all powers on a group of soldiers.

14. *Eshugbayi Eleko* v. *Government of Nigeria* (1931), Ac 662.

15. In Okonkwo, *Introduction to Nigerian Law* (London, New World Press, 1980), 41.

16. Justice R. O. Fawehinmi, guest speaker on the occasion of the Nigeria Bar Association (Ondo) dinner, September 22, 1988. But non-Nigerians, including the chairman of Royal Dutch Shell, derides such rules based on morality and attacks local and international campaigners as "moral imperialists" and "moral relativists" respectively.

17. C. J. Osborne in *Lewis* v. *Bankole* (1908), I NLR 81 at 100-101. Also see *Owonyin* v. *Omotosho* (1961), 1 All NLR 304 at 309.

18. NWLR 182 (1990) at 207. Also see Professor Niki Tobi at page 1 (see footnote 3 above) where

the learned justice affirmed that in the administration of justice in traditional society, customary law ensures "stability of society and maintenance of social equilibrium. The most important objective was to promote communal welfare by reconciling divergent interests of different people."

19. Alan E. Boyle and Michael R. Anderson, *Human Rights Approaches to Environmental Protection* (Life Books, London, 1996), "A Critique of Anthropocentric Rights," 71-87.

20. See "Treaty Between King Pepple of Bonny and the Chiefs of Andony," December 22, 1846, in the Jewjew House or Parliament House, Grand Bonny, in Alan Burns, *A History of Nigeria*.

21. Lord Atkin in Eshugbayi's case.

22. Some parts of it are in limbo, having been suspended or abrogated by the military regime in power.

23. See Decree 107 of 1993, Section 1.

24. Speeches by Nnaemeka Achebe and Alan Detheridge before the Irish Parliament and the European Parliament in January 1996 and also in several public statements.

25. Discussion with ERA activists at the RGS Shell seminar, April 15, 1996.

26. Korokoro, Botem Tai, Bomu, K-Dere, and several other Ogoni villages suffer environmental devastation courtesy of Shell.

27. "Land" here is used in the general sense. It therefore includes everything on it: forests, animals, trees, humankind, stones, minerals, and more.

28. See also the Water Works Law 1915, Cap 131, Laws of Rivers State. (Also applicable in Bayelsa State.)

29. This law prohibits the fouling of water and the "vitiation" of the atmosphere.

30. Western historians choose to call it the "Second World War."

31. Oil was first discovered in Oloibiri in the eastern Delta on August 5, 1956. According to I. F. Nicholson, *The Administration of Nigeria 1900-1960*, 119: "Oil was first discovered in Epe in 1908 by the Egerton colonial government—the oil was thick and not marketable at the time."

32. M. A. Ajomo asserts that "over 50 of such laws operate at the national level." See "An Examination of Federal Environmental Laws in Nigeria," in Ajomo and Adewale (ed.), *Environmental Laws and Sustainable Development*.

33. Since customary laws are recognized and accepted by national law, the trusteeship duty requires the government not to deal improperly with land, especially if it will affect local customs. See Regulation 18 of the Petroleum (Drilling and Production) Regulations.

34. The Agricultural Act, LFN 1990.

35. Chief F. B. Alawari, a traditional ruler in Mini-Abureni, defines the environment as "everything we have here and we see around us. Our forests, ourselves, the fishes, the animals, our gods, stones, the sun, the moon, the stars, everything, everything . . ." Chief Alawari also added, "We have several environmental laws. I can myself enumerate four dozens of such laws for you right now." Interview with F. B. Alawari in Okoroba, January 1994.

36. Here we mean "community forests," which are divided into (1) sacred forests, (2) forests in burial grounds, (3) forests of common use, (4) forests left as boundary demarcation between communities, (5) forests left for future use (fallow), and (6) family heritage forests. Forests here mean: mangroves, freshwater swamp, lowland rain forests, and barrier island forests.

37. Forests forcefully excised from local people and constituted into state-government-controlled reserve. Despite the huge forest estate and high biodiversity, there exists no national park in the Niger Delta.

38. Nigerian Forestry Act 1937, which deals with indiscriminate deforestation resulting from timber exploitation. The law has percolated into state laws.

39. Forest Law 1956, Cap 55 in Vol. iv, 1463.

40. These are all oil-producing states, and Shell operates or has operated in all of them.

41. Cobbina Dede, lawyer and environmentalist in Ozezebiri community in Southern Ijaw Local Government, Bayelsa State, in conversation with Oronto Douglas, January 5, 1995. Dede also said: "Until such Odumu vacates the place, it is the legal owner and community law recognizes such right as just."

42. The companies include Willbross Ltd., among several others.

43. Chief Irene Amangala of Oloibiri in conversation with Oronto Douglas, September 1994.

44. This simile is now in common use in the community. The width of the seismic lines was recently reduced to one meter.

45. Shell currently operates over ninety-four flow stations.

46. The Agip terminal, although Shell uses the facility as well. Shell is the operator of the NNPC, Shell, Agip, and Elf joint venture. The Forcados and Bonny terminals are under the exclusive control of Shell.

47. The Utorugu Gas Plant is in the Ughelli area of Delta State.

48. Professor Bruce Powel, Institute of Biodiversity, University of Science and Technology, Port Harcourt. Professor Powel asserts that loggers from Lagos and elsewhere have turned the Delta into an all-comers game. Loggers pay as little as sixty naira (less than a dollar) for a stand of timber. See also Greenpeace Nederland, *The Niger Delta*.

49. The World Bank, *Defining an Environmental Strategy for the Niger Delta*, May 1995, Vol. 2, 95.

50. This information was given to World Bank officials by workers of the Nigeria National Petroleum Company (NNPC).

51. For example, the gathering of ogbono (*Irvingia gambonensis*), bitter kola, and other nontimber forest products is part and parcel of the people's way of life.

52. Medicine, drinking water, fish from forest ponds in Andoni, Ogoni, Etche, Urhobo, and Abureni have been affected.

53. The oil companies sometimes construct canals to access a well head. The World Bank report

said that "canal construction has precipated some of the most extensive environmental degradation in the region" of the Niger Delta. The destruction of forests, waterways, and land in Okoroba by Shell in 1991 is an example. See p. 34 of World Bank report.

54. Sections 16(2)b and 17(2)d of the 1979 Constitution makes such action unlawful. The sections say the exploitation of resources other than for the common good of the people is to be prevented. Even if the common good is the whole of Nigeria, the provision does not sanction the practice of robbing Peter to pay Paul, which is what Shell's activities in the Niger Delta amount to.

55. Section 234 Criminal Code.

56. Section 257 Criminal Code.

57. Harmful waste (Special Criminal Provisions Decree, 1988).

58. Section 34, FEPA Decree, 1988.

59. The relevant agency is the Department of Petroleum Resources. Reporting to FEPA is mandatory, as the wording of the decree used "and any other" agency.

60. Shell has only just recently incorporated restoration programs as one of the principles to be considered in the award of contracts to oil servicing companies. Nowhere has it been implemented in practice, however. This contrasts sharply with the situation in England, where Shell could commission some fifteen environmental impact assessments and "allay fears and concerns" in fragile ecosystems such as those in Cheshire. See World in Action program, *The Price of Petrol*, Grenada Television, U.K., April 20 and April 27, 1996.

61. In 1996, Shell began working with the London-based Living Earth Foundation, with the support of the newly created Bayelsa State government, to produce "something of a land use plan."

62. Such international organizations as Pro Natura International, Greenpeace, and the World Bank have succinctly chronicled the devastation of the Niger Delta by Shell. In Nigeria, local NGOs and journalists have also brought Shell's activities in the area to the attention of the nation at large.

63. See also Regulation 19, and particularly the comments of Olisa Agbakoba SAN, a leading human rights advocate in Nigeria, in his paper "Legislative Framework for Oil Operations in Nigeria," 1996, 10.

64. Olisa Agbakoba, ibid., 11.

65. Shell holds some 40 percent of the shares in National Oil, the giant petroleum products retailer in Nigeria. With plans by the government to privatize this lucrative sector, Shell's stake is likely to increase.

66. Shell executives are comfortable with this arrangement, as it allows them to do exactly as they wish.

67. We relied on Dr. Uwem Ite of Lancaster University, Dr. Julian Caldecott, Chidi Odinkalu, and Uche Onyeagucha for background material in this section.

68. E. Hutchful, "Oil Companies and Environmental Pollution in Nigeria," in Claude Ake (ed.), *Political Economy of Nigeria* (London: Longmans, 1985). Shell has since confirmed that it is guilty of all

that Hutchful observed in his essay. See *Price of Petrol*, 1996; SPDC, *Nigeria Brief*, 1995: *The Ogoni Issue*, 8 and 9, where the company admitted spilling 1,626,000 gallons of oil, flaring 1,100 million standard cubic feet of gas daily, and causing acid rain one month in the year. Since 1989, Shell also said, it has been recording an average 221 spills per year in its areas of operation, spilling about 7,350 barrels yearly. Shell Nigeria Brief, *The Environment*, May 1995, 3.

69. *The Petroleum Industry and the Environment of the Niger Delta*. Vol 2, proceedings of an NNPC seminar, Lagos, 1979. See particularly A. M. A. Imevbore: "The Impact of Oil Pollution on the Biodata of the Niger Delta," 87–102.

70. A Shell official who frequents the Kalaekule and Opobo South fields confirmed this. Brian Anderson, former chief executive of Shell in Nigeria, also said that the company did not have a waste treatment plant as of May 1996.

71. J. P. Van Dessel, former Head of Environmental Studies at SPDC, commented about this custom: "As the accessibility of these pockets (that is, mangrove forests) is generally low, even by canoe and for 'bushmeat hunters' wildlife can be very interesting (e.g., monkeys, snakes, and birds). Regularly, juju shrines can be found erected along the creek banks: a dead cock tied to a stick, some kolanuts on a plate, a chair and a few bottles of Coke, indicating that also in this respect some parts of the modern world are more influential than others." See Van Dessel, *The Environmental Situation*. See also Updated ERA report, January 1996.

72. This contrasts sharply with the blue flame ERA observed in some fields other than Shell's in Nigeria.

73. "Treaty Between King Pepple of Bonny and the Chiefs of Andony," in Alan Burns, *A History of Nigeria*.

74. E. J. Alagoa, *A History of the Niger Delta: An Interpretation of Ijo Oral Tradition* (Ibadan, University Press, 1972). Also see J. R. L. Allen and J. W. Wells, "Holocene Coral Banks and Subsidence in the Niger Delta," *Journal of Geology*, Vol. 70, No. 4, 1962; and P. A. Talbot, *Tribes of the Niger Delta* (London, Oxford University Press, 1932). Talbot commented: "The Niger Delta therefore with the exception of a few small tribes, occupied by these strange people [the Ijaw]—a survival from the dim past beyond the dawn of history whose language and customs are distinct from those of their neighbors [the Ibos] and without trace of any tradition of a time before they were driven southwards into the regions of somber mangroves."

75. G. O. M. Tasie, *Christian Missionary Enterprise in the Niger Delta 1864-1918* (London, Brown and Johnson, 1978).

76. Alan Burns, ibid., 147.

77. The case of *Owoyin* v. *Omotosho*.

78. Thurnstan Shaw has noted that "where we have written records as in the Niger Delta going back four hundred years, we can see how oral traditions can change according to current political

alignments (Jones 1963, 1965); but for the Yoruba we have no similar time depth of written history to serve as a check on oral tradition."

79. This contrasts sharply with the Judeo-Christian tradition, which gives man unfettered license to "arise, kill, and eat." Ecological animals include earthworms, crickets, and bats. Social animals are mostly domestic and harmless creatures.

80. Several communities in the Sabagirea area of Delta State celebrate an annual festival to commemorate the "conquer of the bush cow" which killed children, provoked hunger, and caused fear in the region.

81. Byron Edwards, as quoted in Alan Burns, ibid., 147.

82. However, the reckless manner in which Decree 11 of 1985 lists animals as endangered species has provoked anger in some communities since some of these animals are not endangered as such.

83. 1945 Cap 85 as applicable in Iko, Akwa Ibom State.

84. 1917 Cap 35 Vol. ii, 951.

85. "Harmful waste" has been defined by the Harmful Waste [Special Criminal Provisions, etc.] Decree No. 42 of 1988, S.1(3), as any "injuries, poisonous, toxic or noxious substance." A. Ibidapo Obe, a law teacher, asserts: "Oil pollutants, because of its hazardous, harmful, and toxic capabilities, may be said to come within the definition . . . of the decree."

86. The use of the word "rampage" is deliberate and accurate. In at least two cases (Biseni and Okoroba) contractors working for Shell mowed down forests and vegetation indiscriminately. In early 1995 a Shell environmental officer brought an endangered bird caught by a contractor to the home of Professor Bruce Powell of the Niger Delta Wetland Centre in Port Harcourt. The creature was imprisoned in a cage.

87. For a comprehensive analysis of these laws see *ERA Handbook to the Niger Delta*. Laws referred to include the African Charter, the Nigerian Constitution (1979), the Criminal Code, the Harmful Waste Decree, and the Environmental Impact Assessment Decree of December 10, 1992.

88. See *Guardian*, Lagos, April 3, 1994.

89. The pollution in Bomu was the subject of litigation, and the Nigerian Court of Appeal found Shell guilty of destroying farmland, polluting streams, and fouling the atmosphere. See the case of *Shell* v. *Farah*, 1995, 3 NWLR, 185.

90. See the case of *Allar Irou* v. *Shell-BP*, Suit No. W/89/71 WARRI H/C 26/11/73, where the judge refused to grant an injunction against Shell despite pollution on land, which also killed fishes in ponds, because Shell is doing an economically important work!

91. Ighobia Creek, Mini-Abureni, where the sacred crocodile sanctuary exists. The people of Idema, Okoroba, and Agrisaba communities revere this creek.

92. Duration could be between 7 and 10 years, as the cases in Biseni and Engenni areas have shown.

93. This is akin to protection accorded certain types of animals by international organizations like

IUCN. In Sangana certain kinds of fish are not to be eaten. In Odeama the octopus is not eaten as food. Some families do not eat certain kinds of fishes for religious reasons in Okoroba.

94. M. A. Ajomo, ibid., 22.

95. Recent pollution incidents in Iko, 1994 and 1995, Nembe; 1994 through 1996; Okoroba, 1993 through 1995; and Ogoni, 1994 and 1995, were not taken to court because the outcome is easily predictable. Besides, the communities do not have the financial resources to do so.

96. The seismic shooter pleads anonymity. Interview with ERA volunteer in Port Harcourt, June 1994.

97. Earth tremors have been reported in Gbaran, Okubie, Oloibiri, and Peremabiri communities, ERA, 1995.

98. Interview with ERA activist in Kokori, Sapele, and Ughelli West oil fields, Shell Western Division, 1994. Those interviewed included Shell employees who did not want their names mentioned, for obvious reasons.

99. Sections 1, 2, 3, 8, and 9. But see Section 16, where the act exempts government-owned ships. Quarre: Does the joint venture agreement between NNPC [a government agency] and Shell make a Shell vessel or ship exempted from the law? Not in the least. See the various regulations already mentioned. A "ship" according to the act is any seagoing vessel—tugboats, barges, etc.

100. World in Action program, *The Price of Petrol*. See also the relevant sections of the Nigerian Criminal Code.

101. See World in Action, *The Price of Petrol*.

102. See also Regulation 6 (only competent persons are to carry out oil activities). Reg. 9 also requires fencing of any dangerous part of a machinery.

103. When ERA raised the issue with J. P. Van Dessel, the then head of Environmental Studies expressed his frustration, saying he had complained to the relevant officials to no avail. For the long-term effects see *Cambridge Waters* v. *Eastern Leather Company* (1994) 2 AC 264.

104. Oil spillage does not occur every second. Gas flaring, in most Shell locations, does. In 1995 the company said it flares only 1,100 million standard cubic feet of gas every day into the atmosphere. See also *Better Britain Campaign*, Silver Jubilee edition, 1994, 21.

105. For example, the Harmful Waste (Special Criminal Provisions) Decree 42 of 1988, Section 1 (3). Also Federal Environmental Protection Agency Decree 1988, Sections 21, 34. See also Section 247 of the Criminal Code and Regulation 42 of the Petroleum [Drilling and Production] Regulations, including the Associated Gas Reinjection Act, Cap 26 LFN 1990.

106. Ibidapo Obe, in "Criminal Liabilities for Damages Caused by Oil Pollution," said this offense falls on all fours with the Common Law offense and suggested the decisions in *Esso Petroleum* v. *South Port Corp* (1984) 2 QB at 182 will apply.

107. Felix O. Tuodolor, then of Niger Delta Human and Environmental Resource Organization. Interview with ERA activists, 1996.

108. Robert Azibaola, Niger Delta Human and Environmental Resource Organization. Personal comments, May 1996. Also see Environmental Rights Action, *Shell in Iko*, July 1995.

109. Section 30 of the Nigerian Constitution (1979).

110. African Charter of Human and Peoples Right, Article 24. Olisa Agbakoba has warned: "The government of Nigeria must not allow multinationals doing business in Nigeria to violate charter provisions with impunity." In "Legal Responsibilities of Shell as an Oil Operator in Nigeria," January 1996.

111. See Environmental Rights Action, *Shell in Iko*.

112. Shell executives addressing the Irish Parliament on January 27, 1996.

113. Article 36 of first schedule to the Petroleum Act. There is a similar provision in Regulation 21 of the Petroleum (Drilling and Production) Regulation.

114. Obasi Ogbonnaya, *The Guardian on Sunday*, Lagos, April 3, 1994.

115. Shell claims credit for infrastructural development aimed at replacing edifices its workers destroyed in the first place in the course of oil exploration and production. In Okoroba the company offered the community 90,000 naira ($900) as "compensation" for destroying the half-built local hospital during its canalization project. See Shell brief on Community Development, 1995.

116. 1868 LR 3 HL 330. Also see J. Chinda and 5 ors v. *SPDC* (1974) 2RSLR 1.

117. MWSJ 61; 87-88.

118. This reechoes the Alar Irou case. But see *Machine Umudje & Anor* v. *Shell-BP* (1975) 9-11 SC.

155. *Edhemowe* v. *Shell-BP*. Unreported Suit No. UHC 12/70 of 29/2/71 Ughelli High Court; *Onyori and Anor* v. *Shell-BP* and another Supra FCA/B/1/82.

119. Per Uwais Ag. CJ. in *Salamotu* v. *Adamu Yola* (1976) NMLR at 115 at page 117, and the Umudje case.

120. See "The Tragedy of Oil Discovery" in *Ogoni: Trials and Travails* (Lagos: Civil Liberties Organization, 1996), 16.

121. See Irish Parliamentary Foreign Affairs Committee video of January 31, 1996.

INDEX

ABOUT THE AUTHORS

Ike Okonta is a writer and journalist. He was part of the editorial team that founded *Tempo,* the underground newspaper, in Lagos, Nigeria, in 1993. *Tempo* later played a critical role in the forced ousting of the dictator General Ibrahim Babangida in August of that year. Okonta also worked closely with the late Ken Saro-Wiwa and other MOSOP activists in Nigeria and is on the management committee of Environmental Rights Action/Friends of the Earth, Nigeria. He is presently at St. Peter's College, Oxford, England, where he is writing a doctoral dissertation on the ongoing social and environmental crises in the Niger Delta. Ike Okonta's first collection of short stories, *The Expert Hunter of Rats,* won the Association of Nigerian Authors Prize in 1998.

Oronto Douglas is Nigeria's leading environmental human rights lawyer. He is deputy director of Environmental Rights Action/Friends of the Earth, Nigeria, and has been a visiting lecturer and speaker at community-organized events, international conferences, and universities all over the world. Douglas was a member of the legal team that represented Ken Saro-Wiwa before he was murdered by the Nigerian military junta in November 1995. He took degrees in law at the University of Science and Technology, Port Harcourt, Nigeria, and De Montfort, Leicester, England, and his articles and speeches have been published in books, journals, and magazines in Nigeria, Europe, and the United States.

Okonta and Douglas are Fellows of the George Bell Institute, Birmingham, England.

ERA: Environmental Rights Action/Friends of the Earth, Nigeria

Environmental Rights Action (ERA) is a Nigerian advocacy nongovernmental organization concerned with the protection of the environment and the democratization of development. ERA is committed to the defense of human ecosystems within the framework of human rights and the promotion of environmentally responsible practices by governments, corporations, and the people. It takes its mandate from Article 24 of the African Charter of Human and Peoples' Rights, which states that: All people have the right to [a] generally satisfactory environment favorable to their development.

ERA is the Nigeria chapter of Friends of the Earth International (FoEI) as well as the coordinating NGO in Africa for Oilwatch International, the global network of groups concerned about the effects of oil on the environment of people who live in oil-producing regions. ERA is the 1998 winner of the Sophie Prize, the international award in environment and development.

E-mail: eraction@infoweb.abs.net